# Carbon 350

*What We Need to Do for a Climate Solution*

ISBN: 979-8-9953406-1-4

Walden Woods Press
12255 Holly Lake Road
Drummond, WI 54832

*For Maria and Vee*

# Contents

# Introduction

The current debate on climate and climate solutions has confused a lot of very fine minds. Good people with good values and deep concerns want to do the right thing. They want to help solve climate change. They want to do their part. They want to contribute to a solution as best they can. The desire is deep and serious. The problem is that they aren't even close to having the impact they could.

Much of this has to do with the way the debate has been framed. The first framing pitted the vast majority of the scientific community against a few outliers funded by the fossil fuel industry, which elevated denialism into a plausible notion—one that captured many people. The petroleum industry funded this starting in the 1970s, even though their own scientists at the time were warning about climate change. That framing was dominant through the 1990s and beyond—in large part finally challenged significantly by Al Gore and *An Inconvenient Truth.*

The framing switched again in 2003 when BP, the global oil and gas company, launched their PR campaign focusing everyone on carbon footprint and proclaiming, "It's a start." It is obvious now that this was another distraction meant to push the public focus off of the industry and onto personal, individual action and accountability. The fact that even activists like Greta Thunberg focus on carbon footprint is a testimony to the incredible effectiveness of this campaign. Today, when you ask most people about what they can do about climate change, the answer will be given in terms of carbon footprint.

These two frameworks—the asserted legitimacy of denialism and focus on consumer accountability through carbon footprint—have controlled how we think about climate change. They are also the primary reasons we are not seeing the results we need fast enough. As noted throughout this book, pre-industrial carbon in the atmosphere was 280 ppm. Scientists say we can withstand 350 ppm and not overheat. As of 2026, we are at 429 ppm. We need to reduce carbon in the atmosphere. Other GHG's must come down too, but carbon is the lead indicator. We know the number. It is carbon 350. And we are still going the wrong direction. Global emissions in 2026 were up another 0.9%, and we know that the current emissions level is far more than the Earth's systems can absorb.

Given the extreme weather events being fed by climate change—especially floods, droughts, fires, and extreme heat—this movement in the wrong direction constitutes an emergency. More and more people see it and feel it. Climate anxiety is now being experienced by large majorities of people, with one study in the *Lancet* showing that among youth and young adults, 84% are afflicted with levels of climate anxiety. Climate denialism is shrinking. Governments and businesses are planning for climate contingencies. Insurance companies are avoiding climate-driven risk. Reality is setting in. Yet there remains little guidance on the key question facing individuals and society: What do we do?

This vacuum has led to a plethora of distracting, meaningless, and downright impossible ideas. Some come out of the carbon footprint framework while others seek to exploit the disruption climate change is causing. One of the main purposes of this book is to debunk these ideas. It is to take on the counter-orthodoxies of degrowth, the Green New Deal, carbon footprint, consumption reduction, and others so that the reader can see through them and get to what really matters in climate: getting us down to carbon 350. Too much energy and focus by well meaning, committed people is going into meaningless, low impact efforts. I will show the reader where the high leverage points are and where each individual can make the biggest impact on solving the climate problem. I will warn you now: It's not about refusing to fly or insisting that everyone give up their cars and ride the bus. We simply cannot get there that way.

In the first section of the book, I hope to dismantle the frameworks and counter orthodoxies that have diverted many fine minds. In the second section, I address the question: Well then what should we do? Here, I entertain the notion that our own ideologies may be blinding us, and to see what we can when the blindfold is removed. It is optimistic to see the most powerful structures in society coming around to support climate solutions—including the market, business, government, and more. I also look at the state of green technologies and their prospects for coming to fruition. Whatever else may happen, we will need ways to create, use, and harness energy that does not generate carbon. We need to understand what is possible today and what will likely be possible tomorrow, and we also need to confront the impacts of deploying those technologies. This section covers solar, wind, battery technologies, and more so that we have a level playing field of feasibility in the technologies available and those that are still developing.

Finally, in section three, I outline an approach to making the biggest difference you can with the limited time, dollars, and effort we each have available. I also map out what society needs to do. It goes beyond the simplistic clarion call: "Leave it in the ground!" which will never work, and I am not satisfied to tinker around the edges of your carbon footprint. Either you take climate change seriously and therefore revise your life, or you don't. Refusing to eat meat, for example, is not a climate strategy. We need to get serious.

In the end, I hope the reader will learn to see through the frames provided by those with conflicted agendas and corrupt interests, especially the oil companies. I hope we can all see through the deceits that have guided our thinking so we can move ahead to impactful action. I hope that once we can see where those high impact actions are, more of us will commit our energies to taking those actions and creating the solutions we actually need.

After nearly two years of research and writing this book, I am convinced of one crucial thing: *There is an answer to climate change, and we can do it.* In fact, we will do it. The answer is getting to carbon 350 ppm or less. The only real question is how much damage will be done before we get there? How much suffering will we endure? How many people will die? How many species will go extinct? Climate change has already created the end of the line for too many people and too many species. Our own well-being as individuals and as a species is now at stake. Clear thought and massive, focused action are the only way. Let's get going.

## SECTION 1

# *Carbon Footprint, Other Delusions, and the Betrayal of Climate Activists*

Carbon footprint has become the mainstay of the climate movement, with individuals thinking about, dwelling on, and even feeling guilty about their own carbon footprint. Most people don't realize that's exactly what the idea was meant to do when it was popularized by none other than BP in the early 2000s. As the website Climate Trade described it: *"The ad encouraged people to calculate their personal carbon footprint – using BP's calculator – to find ways to reduce it, with the slogan "It's a start"."* The ad effectively changed the conversation to focus on personal choices and contributions, while deflecting attention from the enormous emissions of BP and its industry peers. It was advertising and PR genius. Hence, the most commonly heard rallying cry of the climate activist movement—reduce your carbon footprint—was given to us by big oil! And still, activists use these terms without the least sense that maybe we've been had. I hate to say it, but yes… we've been had.

Carbon footprint is just one of the many misguided ideas that seem to have possessed people's minds. In this section of the book, we will examine those ideas, along with carbon footprint, to help eliminate these distractions from the debate. This step is essential because conversations on climate change inevitably get sidetracked into these notions—save the earth, veganism is the answer, or any of a number of other ideas that effectively stop action that will work.

# Why You Can't Control Your Carbon Footprint

In the brief introduction to this section, I described the source of popularity of the phrase *carbon footprint.* It was used to distract people from a focus on the dirtiest industry out there, and onto themselves. The potentially corrupt origin of a term, however, is no reason to reject it if its use is helpful to the cause. *Carbon footprint,* however, is not useful to the cause. Millions of people, thousands of books and articles, most major climate organizations, and hundreds of thousands of businesses are fixated on carbon footprint. This, despite the fact that it cannot be accurately calculated, managing it is completely out of your control, you can't shrink it, the moves you make to manage it are meaningless, and all the energy you put into it distracts you from those actions that could make a real difference. The focus on carbon footprint is not benign. It is stealing the energy humanity needs for solving the climate crisis. We need to break ourselves from the grip of this terrible, inaccurate, and ineffective term.

## *Bogus Carbon Footprint Calculators*

If you have ever thought about your carbon footprint, there is a good chance you tried to calculate it. Perhaps you went to the online calculator BP provided, or perhaps you don't trust them and went to the EPA or the Nature Conservancy calculators. Because, why wouldn't you? If you want to do something about your carbon footprint, the first question is: How big is mine today? And then you ask, and what can I do to shrink it? These are reasonable questions, but there is a big problem.

The guides that measure carbon footprint calculate it on a collective basis, not individual. That's because "carbon footprint calculators" are based on all kinds of factors that have little or nothing to do with your individual actions and choices.

## *Nature Conservancy Calculator*

The calculator by the Nature Conservancy, for example, will change radically based only on changing your location. In my case, I threw in some assumptions and got 41 tons of carbon annually. I changed my location to a town a hundred miles away, and it was suddenly 52 tons annually. Same exact assumptions on actions, expenses, income, spending, driving, and everything else—just a different town.

Doesn't this raise a serious question? How can the exact same lifestyle parameters in a slightly different location yield such different carbon footprint results? If I drive the same car the same distance and speed, why would my carbon footprint shift? If I heat the house exactly the same, using the same amount of natural gas, why would my carbon footprint change? Indeed, the inputs on this were exactly the same, but the result came out different. That means the calculator is using non-entered information to adjust the calculation, and it makes this shift solely on the fact that I changed locations. As a consumer trying to figure out how to reduce my emissions, this suggests that the best thing to do is live in one place rather than the other. That way, without any change in lifestyle, I can achieve a 20% reduction in carbon footprint. Isn't that great?

Except it is obviously false. It gets us thinking about taking actions that will have absolutely zero impact on emissions. Given it's history, it is tempting to think that is exactly what "carbon footprint" was meant to do.

## *UCS Calculator*

The calculator by the Union of Concerned Scientists (UCS) does the same thing, although they are a bit more transparent about it. In this article, they talk about why Wyoming residents have a per capita carbon footprint of 104 tons while New Yorkers are just under 8 tons.

Even more to the point, UCS calculates a table showing the per capita carbon footprint by state. Wyoming residents generate 104 tons of carbon whereas those in states like Connecticut, Oregon, and California are less than 10 tons. It is improbable, at best, that the reason for this difference is that Wyoming residents consume ten times more food, ten times more heating and cooling energy, drive ten times more than people in other states, or fly ten times more than other people. To reach this disparity, they would have to do all these things at ten times the level, but on an individual basis, that doesn't even pass the smile test.

The calculators leave one with a bizarre implication: Live in Wyoming? Your per capita carbon footprint is 104 tons of CO2 per year. Live in New York? It is less than 8 tons. So, if you live in Wyoming and you are concerned about your carbon footprint, what does this imply? The best thing you could do for the planet is to move to New York — your carbon footprint will go down by 92% without any changes in how you live! Wow!

What a bunch of nonsense.

In reality, what this shows is that the way we think about carbon footprint has no impact whatsoever on climate change. The UCS article tries to illustrate the differences between these states, and unsurprisingly, if also unwittingly, illustrates how personal lifestyle choices are nearly irrelevant to calculating the number.

Why is the carbon footprint of New Yorkers so low? According to the article, it is because a big portion of New Yorkers work in financial services, the concentration of people in New York City makes mass transit highly viable, and New York's higher cost of electricity disincentivizes usage. Think about that. If you live in New York, none of those factors are within your personal control. Not one. You might choose to work in financial services, but your carbon footprint is low because so many of your neighbors do so, not because you do. Mass transit is virtually the only way to get around the city so millions of people use it all the time. And when was the last time you got to determine the rate you and your neighbors pay for electricity? Hence, your apparent carbon footprint is reduced because of the choices other people make and the infrastructure around you.

Consider this. If you live in upstate New York where you have all the same challenges as those in Wyoming — rural area, long driving distances, cold weather, mountainous — this calculator suggests that your carbon footprint is radically lower than those who live in Wyoming. It is not because your lifestyle is 92% more efficient than those in Wyoming — it is because your emissions are balanced by those of the more efficient city and those in Wyoming are not. Concerned New Yorkers and concerned Wyoming residents who live the exact same carbon-reducing lifestyles to reduce emissions by 20% in their personal lives appear to be having radically different impacts. One amounts to a reduction of 1.6 tons of carbon, whereas the other is over 20 tons for the exact same lifestyle. If you believe that, I have a bridge to sell you.

### *EPA Calculator*

The calculator by the EPA pretends to ignore any bigger issues by showing just a slice of one's actual carbon footprint—those associated with energy usage, transportation, and waste. Yet it, too, starts by collecting your location and household size, as if those would not be accounted for in the amount you spend on electricity. Maybe they are adjusting for the cost of energy or maybe not. It isn't clear to the casual user.

That said, the calculator provided by EPA is most interesting. It provides not only an estimate of carbon released by heating, cooling, and other uses of energy but also from driving and waste. It also shows you how certain actions can reduce your contribution. That's great, but here's what it does not cover: The average total carbon footprint of Americans based on per capita calculation is just over 20 tons, while the average carbon footprint for Americans from these three categories EPA includes is about 8.5 tons, according to the calculator. *What about the other eleven tons of carbon?* They are not included because nothing in your personal choices will affect them. They are systemic.

This review of carbon calculators illustrates the salient point: While a calculation of carbon per capita can make sense to compare the average between regions, it makes no sense at all for calculating *your personal carbon footprint.* To see just how ludicrous the individual calculation is, let's turn to the most definitive attempt yet to calculate it for individuals—Mike Berners-Lee's book *How Bad Are Bananas?*

# Why You Cannot Calculate Your Carbon Footprint

In comparison to the carbon footprint calculators, a much more honest and complete attempt to calculate carbon footprint appears in Mike Berners-Lee's book *How Bad Are Bananas? The Carbon Footprint of Everything*. Berners-Lee wrote the book apparently to help people manage their carbon footprint. As one would expect, he starts with a definition of the term. "I'm using the word footprint as a metaphor for the total impact that something has. And I'm using the word carbon as shorthand for all the different global warming greenhouse gases."

Then he goes on: "The true carbon footprint of a plastic toy, for example, includes not only direct emissions resulting from the manufacturing process and the transportation of the toy to the store; it also includes a whole host of indirect emissions, such as those caused by the extraction and processing of the oil used to make the plastic in the first place." Berners-Lee further clarifies that this leads to an infinite number of pathways, with but one example being the contribution of the carbon footprint of a paperclip, which would include the steel that goes into it, the mining, the machinery that operates in the mine, the fuel, the refining, and so on *ad infinitum*.

Once seeing this, he makes the single most important statement in the book:

"The carbon footprint, as I have defined it, is *the* climate change metric that we need to be looking at. The problem is that it is also impossible to measure."

The reason has to do with a single important fact—the measure of carbon footprint is enormously complex and involves an infinite number of upstream pieces of information, most of which are not available and none of which are in a consumer's control. On a car, for example, the source and production methods of the steel matters, as does the source of the energy used to produce the steel, and the carbon created to obtain the energy. The same is true for the seats in the vehicles, but also for the leather on the seat, the switches on the seat, the mounts to the floor, the padding in the seat, the wires for the seat warmer if you have one, and on and on. Each of those pieces has its own carbon footprint that is affected by its own processes and suppliers, and each of those by their processes and suppliers as well. Indeed, every part of the vehicle has its own impact, and every part that goes into every part has its own impact, and the assembly of the vehicles has its own impact. And then you start driving it.

This complexity renders the calculation of a carbon footprint for any individual product, service, or action nonsensical. It is utterly meaningless.

Despite his insight into the impossibility of measuring carbon footprint, Berners-Lee argues that an estimated gross level of measurement is good enough… "they will do…" he says. "The level of accuracy that I have described is good enough." But that level of accuracy is about an order of magnitude from reality. He says that 5.5 lbs of carbon dioxide, for example, is "almost certainly between 2 and 20 lbs." Which is to say, "Who knows?" It could be on third as much or four times as much. From the low end of 2 to the high end of 20, the difference is a full order of magnitude! Rather than see the flaw in his logic, he insists that personal choices about carbon are the answer and that grossly erroneous numbers are good enough to make decisions.

He is saying we should all be focused on our carbon footprint, even though we have no way to measure it or know if it is going up or down. Additionally, many causes of its increase or decrease have absolutely nothing to do with our behavior or choices (as the calculators show). Now, if you cannot measure something, you can't manage it either. You have no idea how much of an impact an effort is making, so you can't tell if it is worth it or not. Should I do more? Should I do less? Is this the best thing or is that? There is no possible way to know if you can't measure it, and Berners-Lee says it as clearly as possible—carbon footprint is "impossible to measure."

This notion of carbon footprint and accounting for all the upstream effects provides a kind of intellectual satisfaction to many people, even as it deceives them. The calculators change radically based on factors that have nothing to do with your personal decisions.

The items you might want to calculate carbon footprint for might give you a number within an order of magnitude of correctness, if you are lucky. And many choices you might want to make to reduce carbon footprint have no impact whatsoever unless hundreds, thousands, or millions of other people make the same choice. A good example? Deny yourself that flight to Mexico in February, but unless the airline can't fill the other 285 seats on that flight, the flight goes anyway, and the total impact of your sacrifice on carbon emissions is exactly zero.

Another empty intellectual satisfaction is described by Berners-Lee himself.

"I hope that by the time you have read [this book] you will have gained such a sense of where carbon impacts come from that you will be able to make a reasonable guestimate of the footprint of… everything that you come across. It won't be exact, but I hope you'll at least be able to get the number of zeros right most of the time." Which is to say: *I've done my damnedest. It is impossible to calculate, so hopefully guessing the orders of magnitude will help.*

The argument seems to be that since the differences in different activities appear to be enormous, understanding the comparative enormity of each can help you choose high-impact moves. Here's his example:

"A friend recently asked how he should dry his hands to reduce his carbon footprint—with a paper towel or with an electric hand dryer. The same person flies across the Atlantic literally dozens of times a year. A sense of scale is required here. The flying is tens of thousands of times more important than the hand drying. So *my friend was simply distracting himself from the issue.*" (emphasis added).

This is pretty much the case I have been making on carbon footprint overall—it is a distraction from the really big problems and really big commitments we must make to solve the climate change problem. While Berners-Lee tries to define the order of magnitude differences between different personal activities, he has missed completely the even larger impact of real commitments people can make through decisions that are not measurable at all with carbon footprint.

People who like to use individual lifestyle choices and their impact on carbon footprint nearly always preface their comment with these words: "If everyone did [x], we could reduce carbon by [y]." That passes as good reasoning, but since the predicate—If everyone did [X]—is false to begin with, the implication is as well. They never discount the impact of their supposed result by the number or percentage of people who will never do that, thereby rendering their arguments meaningless. It is the craziest thing that we have let our most important public debate get hijacked by circular reasoning. The circle must be broken.

The problem is that this uncovers a real truth. While it may be nice to think that by not flying you are reducing your carbon footprint, or by not eating meat you are making a difference, or that avoiding a night in a hotel is making a contribution to the climate change fight, you are actually deluding yourself. If you don't buy the ticket on the flight, the flight still goes. If you don't buy the beef, someone else will. The hotel will stand there heated, lighted, and furnished whether you show up or not. Hence, not a single one of these choices *as an individual* has the slightest impact on overall emissions. Yes, you can enter a favorable number into a calculator and smugly see that your carbon footprint is lower according to the calculator, but your actual reduction is zero.

Berners-Lee's calculations provide the service of letting us see this very important fact: *Individual carbon footprint of a product, service, or individual choice is a fallacy and a con-job. Carbon footprint can only be calculated in a meaningful way on a collective basis or at the point source—the tailpipe, the smokestack, or the chimney.* Looking at these direct emission sources, however, can yield some strong results and don't require self-delusion, as we will see.

## *The 10-ton Lifestyle*

Berners-Lee's book was written in 2011 and Berners-Lee indicates that the average American carbon footprint at that time was 24 tons per year. He then promotes the idea of a 10-ton lifestyle, which for Americans would represent a 65% reduction in carbon footprint on average. He claims you divide the ten tons by twelve to get a monthly number and then compare your activities to the impact. "A short car commute, a daily cheeseburger, and some wasteful lighting habits could easily use up a quarter of the 10-ton budget," he says. But wait a minute. If Berners-Lee is right and the actual impact of these activities might be off by as much as an order of magnitude or ten times from the estimated number, doesn't that pose a problem? After all, if one's actual carbon footprint emissions are ten times less than your estimate, these activities use only 2.5% of the annual 10-ton budget, so eliminating them has an almost negligible effect. But if they are ten times more than estimated, you've eliminated your entire carbon footprint for the next 2.5 years and don't need to change anything else at all. There's no way to tell which it is. I ask you: How can that possibly provide you with a way to make a decision?

Plus, there is a huge problem with the 10-ton lifestyle. While Americans would experience that as a reduction in their current lifestyle, most of the world would enjoy it as a huge lifestyle improvement. The ambitions of most societies are toward lifestyle improvements, not toward retrenchment. In other words, they aspire to use more energy and improve their day-to-day lives—and today, that means more carbon. As they do so, the carbon per capita in those societies will also go up, and that will be catastrophic. Berners-Lee even admits this: "If everyone went in for 10-ton living all over the globe, emissions would skyrocket by 40 percent." The 10-ton lifestyle is hardly a solution.

In 2021, the global per capita carbon footprint was 4.69 metric tons. According to Berners-Lee, the UK in 2011 had a goal of getting people to 3 tons per person per year. Globally speaking, a reduction to 3 tons per person annually is a 36% reduction in total emissions. Keep in mind that the International Panel on Climate Change (IPCC) indicates that we need to cut emissions by 76% to prevent global warming from becoming catastrophic and irreversible. To achieve that, the global per capita carbon footprint would have to go from 4.69 metric tons to 1.09 metric tons. Here's what Berners-Lee says about achieving the more modest reduction to a 3-ton lifestyle. "Ultimately, though, it's virtually impossible for an individual in the developed world to get down to a 3-ton lifestyle anytime soon. That kind of cut *requires the whole economy to be made greener.*" (emphasis added). In other words, there is no way to get to 1.09 metric tons without greening the economy.

And that is the point. Here we have the most rigorous attempt I have seen to measure carbon footprint, and the conclusions he makes are clear:

- Meaningful measurement of individual carbon footprint is impossible, so ballpark estimates varying by an order of magnitude or more will have to do.
- There's no way to know for sure if your lifestyle changes are having an impact.
- It is almost certain that those changes are paltry in the big scheme of things.
- The only way we can get where we need to go is by greening the economy, not personal initiatives.

What we will see is that how you impact the system—i.e., by greening the economy—is far more important than how you manage your alleged carbon footprint. Before we get to that, let's move on from carbon footprint as a meaningful measure of anything, and let's dispense with a few other notions that are dominating the conversation.

# The Green New Deal and Why It Won't Save Us

I consider Naomi Klein an excellent journalist and have gained tremendously from her work—especially her books the *Shock Doctrine* and *This Changes Everything*. Her interviews, meetings, reports, events, statistics, and data are always compelling, and usually new-to-me information that must be confronted. For these reasons, I deeply appreciate her work.

On the other hand, *On Fire: The Burning Case for a Green New Deal* is a different kind of book, one in which she is essentially pitching her favorite solution for the climate crisis. What is that solution? A total remake of everything in society that solves virtually all of the social problems—inequality, racism, misogyny, injustice, consumerism, capitalism… a veritable laundry list of progressive complaints—which she and others call the Green New Deal. We've all heard of it. Little has come of the Green New Deal since the book was written an first legislation proposed in 2019, but the thinking underlying it has persisted, and that thinking is the problem.

Klein is making a fatal mistake—the same one so many climate activists are making today. They see the pending climate disaster not as a specific problem to address in specific terms, but rather as the outcome of a corrupt system that must be overturned. She quotes a core goal of the Ocasio-Cortez/Markey Green New Deal resolution: "stopping current, preventing future, and repairing historic oppression of indigenous peoples, communities of color, migrant communities, deindustrialized communities, depopulated rural communities, the poor, low income workers, women, the elderly, the unhoused, people with disabilities, and youth." To which I ask myself: *What about carbon?*

This is the precise problem with a large slice of the climate movement. We are letting every issue we have with capitalism, every injustice, every problem in the world overwhelm the actually critical work of dealing with carbon. As much as people like Klein want to claim that you cannot solve climate change without a total remake of the economy, it's not true. The energy industry will be disrupted, and so will the home appliance, industrial equipment, and auto industries, but there is more to an economy, a society, and a political system than that. Transition the energy sector to 100% renewable energy, then extract and store carbon from the atmosphere, and you solve the problem whether or not you meet all these social justice goals. On the other hand, no one has been able to show convincingly that solving the social justice issues is in any way determinative of a climate change solution.

One of Klein's best works was her exposé of the *Shock Doctrine*, which is the idea that when disasters strike, hyper-conservative forces see those disasters as an opportunity to rape the public sector, privatize, and run away with the wealth. Milton Freidman, the people at the University of Chicago, and many of the old neoconservatives were instrumental in this doctrine. It was applied to everything from Pinochet's Chile to Bush's invasion of Iraq, hurricane Katrina, and the 2008 financial crisis. It is a powerful insight backed up with many details and strong research.

So it is with a certain shock of my own that I see Klein applying the same idea, but from the other side. The thought seems to be: *The earth is on the edge of catastrophe, we are seeing more and more terrible events, so let's put together a whole new theory of being that we can use and apply as if we have a clean slat—as if climate change has already hapened.* It is no different. She and the Green New Deal colleagues are not solving climate change, they are setting up for an economic revolution, and in so doing, *they betray the movement they profess to support.*

The problem is this: you cannot solve climate change by ending racism. You cannot solve climate change by closing the wealth gap. You cannot solve climate change by ensuring equal rights for women. No. You solve climate change by ending carbon emissions and extracting the overage already in the atmosphere and locking it in the ground. You bring atmospheric carbon down to 350 ppm or less. Social solutions, as important as they are, don't do that.

In order to solve climate change, we need to stay focused on the problem at hand—carbon and other GHG emissions. The economic, social, and justice revolutions being advocated distract us from the very pressing problem of reducing atmospheric carbon. While it is certainly true that climate change is a disruptive force to capitalism, we need to solve it whether capitalism survives or not. So the language Klein uses is definitively not helpful:

*"In this time of rising seas and rising fascism, these are the stark choices before us. There are options besides full blown climate barbarism, but given how far down that road we are, there is no point pretending that they are easy. It's going to take a lot more than a carbon tax or cap-and-trade. It's going to take an all out war on* ***pollution and poverty and racism and colonialism and despair*** *all at the same time." (p50-1) (emphasis added)*

Do you see it again? No war on carbon, or methane, or other GHGs. Why not? Because such a war would undermine the ideology of total societal transformation—it would undermine the *opportunity* of climate change.

Commenting on a 2018 NY Times piece by Nathaniel Rich in which Rich makes the spurious claim that human nature was the cause of not acting on what the science knew in the 1980s, Klein puts the blame elsewhere:

*"Yep, you and me. Not, according to Rich, the fossil fuel companies who sat in on every major policy meeting described in the piece. (Imagine tobacco executives being repeatedly invited by the US government to come up with policies to ban smoking. When those meetings failed to yield anything substantive, would we conclude that the reason was that humans just want to die? Might we perhaps determine instead that the political system is corrupt and busted?)" (p245)*

The comparison between fossil fuel and tobacco is a red herring at best, and I can't help but conclude it is meant to mislead. Tobacco, after all, serves no pressing public need. Energy does. Fossil fuel companies are corrupt in many ways and they hid their own data and funded anti-information campaigns on climate—they were awful in that way. At the same time, they also serve enormous human needs—to heat our homes, move our vehicles, keep industry running, deliver our food, and so on. Energy is at the very center of modern economics, by which I mean not Wall Street, but the actual production and distribution of goods. The conclusion that everything just had to be conspiracy and corruption even though we had, in the 1980s, no viable alternatives for heat, transportation, or manufacturing, exposes the commitment to an ideological position not that different from the ideological commitment of the free market fundamentalists Klein so rightly loathes.

The problem with blame is that it lets us all off the hook. Blame the oil companies, blame capitalism, blame the CEOs, blame corruption, blah, blah, blah. The implication is that when those who we blame correct their behavior, everything will be okay. Therefore, we need to overhaul the system that puts those people in place. Suddenly, the analysis goes to all the other societal ills, and the mind goes numb—we can hardly fathom that other world, and so we just throw up our hands in despair.

In another section of the book, Klein discusses the perspective of Brad Werner, a scientist presenting at the 2012 meeting of the American Geophysical Union. Werner tried to assess the forces at work driving us toward climate Armageddon. His measuring determines that the forces of the dominant economy are hurtling us toward an inevitably dark future, while the primary countervailing forces are those of social resistance, because they may be the only force that can actually slow it down. *"... he is saying that his research shows that our entire economic paradigm is a threat to ecological stability. And, indeed, that challenging this economic paradigm, through mass movement counterpressure, is humanity's best shot at avoiding catastrophe." (p113)*

If so, God help us.

I say that because mass movement counterpressure almost never wins anything, and one of the reasons is precisely what Klein and the Green New Deal advocates are doing—hijacking a specific, definable, winnable issue that needs real attention and fast, and complicating it by projecting their own pet projects onto it in the hopes of getting a bigger or different win for those pet projects. In doing so, they steal the energy and the focus of the real movement. Hence, even when a movement does win, there is always a question—what did we win? No one knows.

Klein claims that we need an all out, full society commitment, but she focuses that commitment on an impossible task—overturning capitalism and replacing it with something that enormous numbers of people will resist. No doubt, we need the all out, full society commitment, but I would hold that the direction and goal need to be different. She illustrates what such a commitment would look like by referencing the Great Depression, World War II, and the nuclear threat of the 1950s:

*"That, most fundamentally, is why the historical precedents from the 1930s through to the 1950s are still useful. They remind us that another approach to profound crisis was always possible and still is today. Faced with the collective emergencies that punctuated those decades, the response was to enlist entire societies, from individual consumers to workers to large manufacturers to every level of government, in deep transition with clear common goals.*

*"Past problem solvers did not look for a single "silver bullet" or "killer app"; nor did they tinker and wait for the market to trickle down fixes for them. In each instance, government deployed a barrage of robust policy tools… all at once." (p 38-9)*

The key to this success is not, as Klein seems to think, the full commitment. It is those last words of the first quoted paragraph: "clear common goals." It should be self evident to anyone living at this time of history, and is obviously self evident to Klein based on her commentary, that we do not have clear common goals around racism, misogyny, inequality, and all of the remaining laundry list of progressive issues she and the New Green Deal people are advocating. The consensus just isn't there. Without it, there cannot be an all out, full society commitment.

You see, for all the other good that may have gone along with the original New Deal, there was one clear common goal: Put people back to work. Whatever else the war may have wrought, there was one clear goal: Beat the Nazis and the Japanese armies. Any other goods that fell out of those efforts were incidental, not causal. A laundry list of benefits may have resulted, but those benefits were not the goal. The goal was simple, clear, and everyone knew what it was.

We can also do this on climate. There is one goal: Reduce carbon to 350 ppm in the atmosphere. That means two things: Eliminate emissions and reduce atmospheric carbon. It's really just that simple. The "full society" commitment then, like in the 1930s and 1940s, should be organized around *everyone understanding and contributing, to the best of their ability and in their circumstances, to make the difference they can make.* Some will invent new technology. Some will reduce their home use of energy. Some will overhaul a business's operations. Some will install rooftop solar, and some will develop windfarms. Some will plant trees to reforest. Everyone can be called to greater and greater contributions, but the question we will all ask each other is this: What are you doing to help? And the societal ethic will be one of indignation if you aren't helping enough. This is how you get a real transformation. Set a clear, understandable goal and orient everyone to that goal.

# The Case Against Socialism as a Solution to Climate Change

In a corollary to the Green New Deal, many people are concerned about "…too much emphasis on technological solutions to climate change mitigation instead of cultural, political and economic institutional change," as one friend put it in a private email. While the emphasis on technology is a criticism worth considering, the prescribed focus on "cultural, political, and economic institutional change" is a deeply dangerous direction. I'll explain why in just a minute, but first, let's look at what this means.

*Why capitalism gets blamed*

This focus on institutional change has come about because of the inherent criticism of capitalism that climate change implies. In a different book called *This Changes Everything*, Naomi Klein makes one of the strongest cases for that inherent criticism, illustrating that even the "free market fundamentalists," as she calls them, understand the implied criticism. Indeed, she argues their clear understanding of that problem is precisely the reason for climate denialism—if climate change is real, it blows a huge hole in free market, capitalist ideology.

Now, if we also look at society, we can see a hoard of social problems, most of which can be traced to capitalism—concentration of wealth, racism (and its connection to slavery and colonialism), misogyny, social injustice throughout the system, criminalization of life, etc. Much of this is based on capitalism's penchant for hierarchy, extractive/exploitative processes, coercion, and the creation of scarcity largely through market mechanisms. Indeed, almost any criticism of capitalism sees these connections.

But here is the thing: Most of these same problems also exist under the only alternative anyone ever points to—socialism. Sure, you could point to an ideal implementation of socialism and argue that it doesn't have to be that way, but we could do the same with capitalism.

Ayn Rand versus Karl Marx. That's a fun intellectual exercise, but at most there are only one or two examples of the pure version of either, and I would argue none at all. Either way, the pure case does not present a practical solution because the instances of all the other cases illustrate that such outcomes are unrealistic idealizations. When we are dealing with something as dire and urgent as climate change, a certain practicality is in order. Our solutions need to work.

### *The real problem…*

The problem is simple—we mischaracterized the underlying systemic causes of climate change. In actuality, it is not *capitalism*, it is *industrialization*—fossil fuel-burning industrialization. Industrialization was only possible with the enormous energy release caused by burning fossil fuels, and some aspects of industrial society still rely on that energy. There are no industrialized societies that are not burning fossil fuels. If industrialization is the problem, we must realize that socialism is also industrialized. Hence, a switch from capitalism to socialism does *nothing* to the underlying cause of climate change. Hence, the "cultural, political and economic institutional changes" that so many people promote as solutions to climate change will not, in and of themselves, lead to the change needed. There is nothing in socialism that is intrinsically a climate solution. There is nothing in institutional change of any type that is intrinsically a climate solution. And if we continue to think that such change is the only way to solve climate change, we put ourselves into a box from which we cannot escape. Moreover, if we allow ourselves to get distracted with institutional change, we will not only lose momentum and precious time, but we could win the institutional change without having any effect on climate whatsoever.

### *The dangerous cost of distraction*

This, then, is why the focus on "cultural, political and economic institutional change"—or the change to socialism—is so dangerous. It delays the crucial work that must be done to solve the climate crisis. As I have said, the climate change solution we need comes down to one thing--reduce the amount of carbon in the atmosphere, nothing more and nothing less. That means emit less and extract as much as possible. If we could focus on these things we can beat this thing.

As we have seen, today, we are at 417 ppm of carbon in the atmosphere, while the UN IPCC says we need to be at 350 ppm or less. As the concentration of atmospheric carbon increases, the world is getting worse on every environmental measure. Unlike other correlations, this is a known causality. The increase in atmospheric carbon is leading to atmospheric warming, and these lead to extreme events from the flood in Pakistan to the Canadian heat dome of 2021, the European heat wave of 2022, the global drought conditions drying up rivers, and a long list of other extreme climate events in 2023, 2024, and 2025. The warming is also changing ecological systems, including their productivity of the foodstuffs on which humanity depends. We can also see that it is going to get much, much worse unless atmospheric carbon can be reduced quickly.

Here's the danger: While this emergency is unfolding, the people who care and want to make a real difference are dithering about institutional change. They are dreaming about solving racism and social injustice. They are advocating socialism instead of deploying clean tech. They are advocating programs like degrowth as if degrowth were a magic bullet even though, as we will see, degrowth has nothing to do with carbon per se.

You see, climate change is *the urgent, existential issue of our time.* Nothing else compares. Hence, advocates of all other issues that can't *actually* compare are attaching their issues to the climate change bandwagon, thereby confusing the issues, distracting attention, and draining energy. *This must stop.* Remember, you can stop racism without doing a single thing to solve climate change. You can bring down the billionaires and heal the hideous inequality in wealth accumulation, and not do one thing to solve climate change. You can turn every economy into a socialist one without making one iota of change in carbon emissions, and therefore climate change. At the same time, we can solve the climate problem without solving racism, inequality, or becoming socialist. We need to reduce carbon in the atmosphere. That means emit less and extract as much as possible. The idea is simple. The execution is very difficult. Difficult enough without the distractions of systemic change, overthrowing capitalism, undoing the banking system, and solving every favorite social problem under the sun. Carbon, carbon, carbon. It's the only thing that matters.

# Market Realities Undercut Socialism as a Carbon Solution

Carbon is produced when we meet the core needs of society. In short, we can't live without it unless we have a replacement. Many of the same progressives with the big institutional change vision I've been outlining also turn their ire on big oil as corrupt, self-serving, and dirty. Indeed, they are not wrong, but there is another dynamic these activist's calls for "leave it in the ground" don't account for—*necessity*. The necessity to heat your home in winter and cool it in the summer. The necessity to obtain food on the grocery store shelf that is shipped there by trucks and grown with the use of energy. The necessity of cooking your food, turning on your lights, obtaining needed products, producing products people need, and making social connections. The necessity of getting to work, getting to church, getting to school, and getting to friends. We've so far been unable to get ourselves off of fossil fuels largely because we have had no alternative way of meeting these needs. Now that alternatives are becoming available, the corruption and power of big oil comes into the limelight.

Reducing carbon emissions is hard because modern industrial society has been dependent on burning fossil fuels to meet society's urgent and pressing needs, whether capitalist or socialist. We need to find other ways of meeting those needs. For the first time, we have replacements, and we all know what they are—solar panels, wind turbines, and geothermal systems for energy; electric vehicles, electric appliances, and electric industrial machinery powered by these renewable systems; and improving efficiency in all of them to produce more with less and to reduce overall need. Some technologies need to get much better, especially electric power storage technology—i.e., batteries. The challenge now in solving climate change is getting everyone to adopt these new technologies, for meeting real human needs without burning carbon is the key to reducing emissions.

### *Green tech or global catastrophe… that's our choice*

The only alternative to global climate catastrophe is the ubiquitous adoption of green energy technology and the appliances and equipment that use it. This means the technology needs to be developed, produced, and costs brought down so that more and more people, businesses, and countries can adopt it. And so I ask the critical question: Which institutional system is better at developing solutions, producing them, and bringing them to market for wide-scale ubiquitous adoption—capitalism or socialism?

You see, socialism is necessary for adoption only when the cost of the socially sanctioned technology is higher than that available in the capitalist market. This is because socialism enables government to mandate such adoption. People don't, on a mass scale, choose to pay more for things if they don't have to. If the solutions are more expensive, then socialism is needed to force that climate-friendly decision.

However, when the capitalist market turns green energy systems and the related high efficiency electric appliances into the low-cost alternatives, as is happening right now, the need for mandates evaporates because individuals and businesses *choose* to adopt the lower cost alternatives of their own volition. There is no need for a mandate because the economic rationality of the capitalist system guides people to the right decision without it. In the conditions where climate-friendly options are less expensive than the others, the institutions and methods of capitalism actually serve the adoption of green energy tech and electric equipment/appliances. So let's be careful what we are trying to overthrow—at least, if we want to solve climate change.

If there is to be an overthrow in paradigm, it is not to replace capitalism, but to overthrow fossil fuel-based industrialism. Our current industrial paradigm is to acquire and burn cheap fossil fuel, releasing tremendous energy, and therefore powering the world to meet a wide range of human needs, wants, and desires. For all the problems of industrialism, its ability to release such energy has resulted in a massive expansion of humanity on earth, astonishing changes in living standards, and incredible amounts of pollution. All are true. Assuming no one wants to volunteer to be part of the seven billion people who would have to disappear to get back to pre-industrial population levels, we have no other choice but to fix the fossil fuel basis of industrialism. And we are doing exactly that with the development and deployment of green energy. Capitalism is now helping us do so.

This is not in any way to argue that capitalism is ideal. Socialism may well be a better way to manage a society, or not. As we stand at the precipice of climate disaster, we have an absolute need to develop and deploy the needed technologies as fast as possible, and this has implications in regard to socialism and capitalism. First, we cannot wait to deploy until capitalism is overturned and socialism installed in the major countries of the world. That will take a very long time and may not ever happen at all. Second, the energy going into the work for such change is wasted at a time when we need it focused on carbon. It is a major distraction and takes the energy of very intelligent, committed, and dedicated people in an alarmingly useless direction. Third, like it or not, we have gotten ourselves to a point where capitalism is a better vehicle for developing and deploying the needed technologies and equipment than socialist mandates. I know this skewers the progressive dream of using climate change for systemic change, but we cling to that dream at our collective peril.

We must realize that we don't have time to wait for institutional change that is unlikely to come about anyway. Climate is a serious crisis. The system we are under—capitalism or socialism—does not matter. What matters is the rapid deployment of green energy technology and the equipment to use it. Let's stop dithering about in the contest between socialism and capitalism; let's use whatever system we have to get the technology deployed. It is, in fact, our only way forward.

# Why Veganism Won't Reduce Your Carbon Footprint

If you want to have an impact on climate change, don't start with vegetables.

The arguments over veganism as a solution to climate change have become nothing short of absurd. The most quoted source I have found on this is the WorldWatch Institute report that says over 50% of global greenhouse gas emissions can be attributed to animal agriculture. Here's one example:

*"A widely publicised report published by the Worldwatch Institute indicates that 51 per cent of worldwide greenhouse-gas emissions can be attributed to animal agriculture, specifically to cattle, buffalo, sheep, goats, camels and pigs which are raised and killed for food."*

The article goes on:

*"Study after study has shown that there's a* **strong correlation** *between climate change and the production of animal-based foods." (emphasis added)*

The problem is there is also a strong correlation between the consumption of whiskey and reading the Bible, but it doesn't mean one causes the other. Nonetheless, the article goes on to this claim:

*"...a global shift towards a vegan diet is vital to protecting the world from the worst impacts of climate change."*

These kinds of reports become amplified all around the Internet and elsewhere, until they become accepted fact. For example, this article on Medium mindlessly claims "*Eating more plant-based meals is the fastest way to shrink your carbon footprint,"* and offers no evidence at all.

### *Conflicting Studies*

Let's use our brain for a minute. The 51 percent claim is preposterous. Sure there are a lot of animals, and no doubt much deforestation occurs as a way to inexpensively produce meat. It must contribute something to the problem. But 51%? Does anyone really think that becoming vegan would reduce greenhouse gas emissions by 51%?

More than electricity generation, all gasoline burned in vehicles, all steel and cement production, and all other industrial activities *combined*?

More credible sources say no. According to the EPA, transportation (29%), electricity generation (25%), and industry (23%) combine to account for over three-quarters of US greenhouse gas emissions. Agriculture only accounts for 10% of US greenhouse gas emissions, and only a portion of that is from livestock. In other words, even if the entire country became vegan — an unlikely scenario — the total emissions reduction would only be 3% or so. Hence, the impact on greenhouse gas emissions will also be quite small.

### *Where Did We Get The Idea Veganism Can Solve Climate Change?*

The original story emphasizing the GHG contribution of livestock came from the United Nations Food and Agriculture Organization (FAO). FAO published a study authored by Henning Steinfeld in 2006, which claimed that livestock produced 18% of global GHG and concluded that livestock was producing more GHG than the entire transportation sector. Although it is a mystery how WorldWatch inflated that to 51% three years later, the claim in the FAO study was eye-catching and got a lot of press.

But there was a slight problem.

### *A Mistake in the Science*

Frank M. Mitloehner, Professor of Animal Science and Air Quality Extension Specialist, University of California, Davis, reviewed the report and raised an issue with the carbon footprint comparison at the center of its conclusion. Here's what he wrote:

*"The problem was that FAO analysts used a comprehensive life-cycle assessment to study the climate impact of livestock but a different method when they analyzed transportation.*

*"For livestock, they considered every factor associated with producing meat. This included emissions from fertilizer production, converting land from forests to pastures, growing feed, and direct emissions from animals (belching and manure) from birth to death.*

*"However, when they looked at transportation's carbon footprint, they ignored impacts on the climate from manufacturing vehicle materials and parts, assembling vehicles, and maintaining roads, bridges, and airports. Instead, they only considered the exhaust emitted by finished cars, trucks, trains, and planes. As a result, the FAO's comparison of greenhouse gas emissions from livestock to those from transportation was greatly distorted."*

Distortion is one thing. What the mistake did was mess up the math such that the 18% of global emissions attributed to animal agriculture cited in the report was off by a large margin.

*The study author corrects his mistake*

This oversight was pointed out by Professor Mitloehner, and Mr. Steinfeld, the original author of the FAO report, corrected this mistake. A subsequent article by him addresses the issue directly:

*"The comparison measures* ***direct*** *emissions from transport against both* ***direct and indirect*** *emissions from livestock. The* ***Intergovernmental Panel on Climate Change*** *(IPCC) identifies and monitors human activities responsible for climate change and reports direct emissions by sectors. The IPCC estimates that direct emissions from transport (road, air, rail, and maritime) account for* ***6.9 gigatons per year, about 14% of all emissions from human activities****. These emissions mainly consist of carbon dioxide and nitrous oxide from fuel combustion. By comparison,* ***direct emissions*** *from livestock account for* ***2.3 gigatons of CO2 equivalent, or 5% of the total****. They consist of methane and nitrous oxide from rumen digestion and manure management."*

In other words, if you compare the same types of contributions to GHG, namely the direct emissions, transportation is 14%, and livestock is 5%--*not 18% as the original paper claimed.* There is no data on the indirect emissions from transportation, so a direct comparison cannot be made.

The point is that *animal agriculture only accounts for a very small portion of total greenhouse gas emissions.* It is not even close to 51%. Science is a process of continuing to peel back layers of the onion, so to speak, until you get to the truth. Big claims derived from the science by ideologically-driven think tanks like WorldWatch Institute must be regarded with skepticism. Eventually, the truth comes out. But sometimes, as in this case, it comes out way too late, and the initial incorrect conclusions can taint public thinking for a long time.

The claim that livestock produce excessive greenhouse gases (GHG) was a mistake. Pro-vegan folks may not like to hear it, but going vegan will have a minimal impact as a climate change solution. Nonetheless, for years — and even today — climate advocates have been pouring energy into this mistaken priority. I've heard from many concerned climate activists that they have gone vegan to "do my part" for climate. This is how these deceptions and self-deceptions play out. Very committed activists put their energy into making themselves vegans rather than doing the hard work of the truly systemic change the world needs.

Before we go on to the next fallacy, let's consider the key ingredients here. First, you have a focus on carbon footprint, a meaningless, unmeasurable number popularized by one of the biggest oil companies in the world. Then, there was mistaken science by a UN agency, the FAO. Third, the mistake was magnified (attributing 51% instead of the original 18%), and the result amplified throughout the climate change community. Now, there are campaigns and organizations dedicated to the idea of getting everyone to go vegan, even though the original scientist says the only data they have would show a 5% impact at the most—and that only if animal agriculture were eliminated entirely.

# Degrowth Insanity is Ecological Suicide

Degrowth, which is also known as steady state economy, is another of our massive diversions from climate solutions in which it seems that the underlying goal is disconnected from actual solutions to climate change. It is an idea for a new economic paradigm with roots in some fairly strong academic work, but as it has come into popular parlance, the message has once again become confused and misappropriated.

## *What Is Degrowth?*

As I will illustrate shortly, degrowth seems to mean different things to different people. Hence, to speak about it, one must carefully define what we are really talking about. I went to the most prominent sites and advocates I could find to try to understand what this is about. There are three primary inputs to help us get a handle on this.

First, the website degrowth.org sells the book The Case for Degrowth and includes this description:

*"Degrowth is not simply a contraction of the economy, it is living meaningfully, enjoying simple pleasures, working less, and sharing and relating more with others and working less, in more equal societies. Its goal is to purposefully slow things down in order to minimize harm to humans and earth systems."*

From this, one can gather a couple of things. First, degrowth is about contracting the economy. It seeks to make the economy smaller. There are words for making economies smaller. If you want it a little bit smaller, it is *recession*. If want it to get a lot smaller, it's a *depression*. Nothing is defined by degrowth as to the size and scope of this shrinkage, but if you have lived through a recession or depression, you would probably prefer to avoid it in the future. Yet according to its advocates, degrowth is supposed to inspire most of the world's population. That's hard to see.

Second, degrowth is about personal lifestyle — how you live. It is "living meaningfully, enjoying simple pleasures, working less, and sharing and relating more with others and working less, in more equal societies." What happens if those aren't things you value? What if championship-level competition is more important to you than working less? What if scientific achievement is more important to you than simple pleasures? What if you just love to drive jet skis fast? I guess proponents would say, "move on to something more meaningful, simple pleasures, and more equal." Which is to say, "Change your life so it is more like mine." No wonder people don't sign up.

A separate proponent of degrowth, The Center for the Advancement of the Steady State Economy, published its policy priorities, which you can find here. There are fifteen "priorities" and they address population control, changing the monetary system, income inequality, resource distribution, the workday, banking regulations, and much more. In places, it reads like a socialist manifesto. But not one single word about not burning fossil fuels. Not one word about carbon emissions.

In fact, the closest this degrowth approach comes to addressing climate is priority number 6: "Employ cap-auction-trade systems in the commons sector for allocating basic resources. Set caps based on biophysical limits. Use auctions to distribute rights to extract resources. Equitably redistribute auction payments through public trusts. Implement a trading system for extraction rights to achieve efficient allocation of resources to those uses with the highest demand."

Read this carefully. A "cap-auction-trade" system assumes that oil, natural gas, coal, and other fossil fuels are going to be capped, auctioned, and traded. And therefore, burned. It says nothing about renewables. Earth-to-degrowth-advocates: Fossil fuel burning is the climate change problem. Simply changing who burns it through an auction-trade system doesn't change the fact of it being burned.

In this incarnation, degrowth appears to be a economic vision reflecting a complete overhaul of the capitalist economy. Such an overhaul may or may not be a necessity based on your perspective, but clearly neither of these versions reflect a primary concern with climate change, per se.

In 1989 two pioneers of degrowth, Herman E. Daly and John B. Cobb, Jr., made their case in a book called For the Common Good: Redirecting the Economy Toward Community, The Environment, and a Sustainable Future. In that book, they had their own list of policy recommendations, many of which focused on tax code tweaks to adjust incentives. The book's underlying assumption is to leave capitalism in place — we'll just try to control it better through politics. I found reading the book many years ago exciting and compelling until I got to the tax code solutions, which essentially convert the economic program into a political program that pretty much anyone could see was a nice aspiration, and unlikely to ever occur. Now, in 2023, the idea has had over thirty years to take root. It hasn't happened. It is pretty hard to see how the next thirty years would provide any different result, and while we wait for appropriate action, climate change is demonstrably worse—much worse.

These three sources all reflect the same fundamental truth—*degrowth is not about climate change.* Degrowth is a vision for a new economy which may or may not have its merits. It is about fairness, distribution, living differently, how much we work, and managing resources. Tangentially, these things will likely impact climate change, and almost for sure, climate change will impact them.

But in reality, it is worse. Because of the priorities of degrowth, you could achieve all the policy goals and initiatives stated and never solve climate change. Degrowth policies leverage capitalism and always leave us on the continuum between capitalism and socialism. While proponents think they are changing the system, the actual recommendations either tweak around the edges with changes to the tax code or they go deeper and advocate a socialist system with high-level control over things like income equality, the number of children you can have, etc. Meanwhile, the global nature of the economy, the industrialism, the corporatization, and the extensive burning of fossil fuels do not change.

Hence, the problem with degrowth is that it actually *advocates the perpetuation of industrial capitalism and the regulatory framework around it* rather than focusing on qualitative change. The large socialist countries of the world have not reduced carbon emissions any more appreciably than the capitalist ones. Degrowth attempts to control and regulate, not **replace** the capitalistic system. The more it does, the more it looks like socialism.

The success of a socialist system depends on who is in control of it. Many degrowth advocates seem to assume that it would be democratically controlled. Even if true, think about our current political situation — do you really want the opposing side of the American political spectrum legislating how many children you can have, limiting your income, or telling you what you can and cannot do with your money? Do you want them to be able to determine your workweek — or that of your employees?

Although some people find a way to trust such authorities, many — perhaps most in the US — do not. And since they do not, they will never go along with such a degrowth strategy. Hence, if degrowth is the answer to climate change, we have no answer at all.

### *The Burden Shifts to Individuals—Again*

The actual academic and philosophical roots of degrowth focus on the economic system and advocate for a political change to its advocates' liking. But as in other cases, that's not how it plays out in the world. Many people have translated degrowth into a call for changes to individual consumer behavior, and it comes off as a caustic blaming of people for going about their normal lives. One commenter, for example, raged about people idling their cars at restaurant drive-thrus in 2020 and 2021 — never mind that that was the only way one could purchase from those restaurants during COVID. It all comes across as blaming other people. "That person is doing something I hate doing, so they are to blame." The change required amounts to an argument that *they should all be like me.* Fortunately or not, most people like being themselves. Because of that, they will not change. And that's why relying on this approach to solve climate change is suicidal for humanity.

But there is another reason, too. The focus on individual behavior is disingenuous. Proponents call on people to "use less," but there are no real numbers, and for most of us, there is no real way to do it.

Let's take an example. Let's say I live in a suburb on one side of a metropolitan area. My spouse works nearby, and my kids and school are nearby — all within 7 miles. But my work, which I enjoy and pays well, is on the other side of the metropolitan area — a 42-mile drive.

So first, in the context of this amount of driving, the call to stop idling my car for five minutes when I pick up a burger (maybe once a week) is ridiculous. That's a solution for climate change?

Second, I hear the call to "use less" as it is legitimately intended and translate it to "burn less gas by driving less." Okay, that makes sense. But, how am I supposed to achieve this? Relocate my family? Even if I did, every mile closer to work is more miles driving back to school, church, and my spouse's work. Should I get a different job? That might be a big sacrifice for me. Is anyone else doing it? I could go to an electric vehicle (EV), but the same degrowth advocates will tell me there is no technological solution and that EVs only achieve fuel shifting. Still others tell me to get on the bus or train and change my commute from an average of 50 minutes to an average of 1.5-2 hours. At this point, the "use less" mantra championed by degrowth advocates creates a vision of making my life worse, not better, enabling capitalism to further erode my family connection since I am not present as much (due to commute time), further stretching the inequities between the haves and the have nots (for the haves will not give up their vehicles), and guilts me for going through a drive-thru. Yet almost nothing has changed on climate.

If I am one who accepts the science of climate change and wants to participate in a solution, I am up against a wall. My short-term needs to eat, heat my home, pay my bills, feed the family, and obtain the goods of life overwhelm the long-term requirements to reduce overall consumption for the good of the planet. There is no way for me to see it other than that my life is going to get much worse to participate in this solution. And so, if given a choice, I won't do it.

And if people won't do it, you have no solution.

This is why, by itself, degrowth is an insane, suicidal strategy. Degrowth advocates, like the Green New Deal and socialism advocates, have hijacked the climate movement to energize their own, largely unrelated, agenda. But degrowth advocates have also managed to transform their ideology into an unattractive guilt trip for people who just want to live their lives while contributing to a solution for climate change. If billions of people on the planet simply will not do what you are proposing, it won't work. The planet heats up. We get runaway climate change.

# Consumption Reduction and the Con Job of Personal Practices

Will consuming less save the planet? Unfortunately, the answer is no, but that has not stopped the ideology of reduced consumption from becoming another failed climate change strategy. The mantra can be heard over and over again in the debate: "We must consume less." "Most Americans can change their lifestyle to reduce consumption by 50% and they wouldn't even feel it." "If we don't consume less, we are f*cked." These are real quotes from comments received. Here's the problem.

The prescription to "consume less" isn't very clear. It is almost always a generalized mantra. "Use less. Consume less." Consume less than what? How much less? If we consume, for example, 10% less gasoline, is that enough on climate? How about 30%? If reduction means the businesses that build things just build fewer of them, is that sufficient on climate? I don't think so. We need to use our very best talents and skills in concert with other people to radically change how things are made and how we procure things, not just how much we procure and consume.

But to try to get at this number of how much we must reduce consumption, I went on a search. It seems that we have an answer from UN IPCC. The first primary objective is to get to a net zero state of emissions. Certainly, that is the main goal articulated in the Paris Accords. A UN webpage says this: "To keep global warming to no more than 1.5°C — as called for in the Paris Agreement — emissions need to be reduced by 45% by 2030 and reach net zero by 2050." The same report goes on to say: "More than 70 countries, including the biggest polluters — China, the United States, and the European Union — have set a net-zero target, covering about 76% of global emissions." In other words, the target is a 76% reduction in emissions to get us to net zero. Those who advocate consumption reduction as a solution, therefore, must have in mind a 76% reduction in personal consumption. That's a lot, so I got to thinking: how much of my consumption do I actually have control over?

*Let's see… I can buy fewer groceries but I already have that pretty well attuned to my hunger needs. I can drive less, but I still need to get to work every day, get the kids to practice, and get to the store for groceries and home repair supplies. I guess I could go out to eat less, but the local restaurants need my business and, dang it, sometimes I am just too tired and can really enjoy the respite going out offers. Out of guilt, I replaced my gas-powered lawn mower with a push mower, and from the standpoint of noise it is an improvement, but I cut the lawn less often because it takes more time. I guess I do use less gasoline in the mower though. I consume by eating, driving, working, heating my home, and taking care of things. Can I eat 76% less than I do today? Heat the house 76% less? Drive to work 76% less?*

My point is that most people think of their consumption this way. They don't see it as extravagant, over-the-top, or massively abusing. They see it as normal and much of it as necessary. Even folks with different values than mine — those who love their four-wheelers and Jet-Skis, for example, or those who feel the need to fly for a vacation in Central America — tend to see their own lives this way. Since life does not feel extravagant, what are they supposed to consume less of? And how much less?

And most important, does it make a difference?

This is the reason why climate change is so challenging. In many quarters, it has been turned into an assault on *personal lifestyle practices* rather than the *industrial systems* that cause it. Here are some examples:

- We are implored to use mass transit and park the car when there would be little wrong, from a climate perspective, with driving an electric vehicle powered by solar energy. Not driving is our personal change, the kind of cars we have is a systemic one.
- We are told not to travel by air because of its carbon footprint, but the problem isn't the footprint, it is the way the planes are made and operated. We have no personal control over how airplanes work.
- Electricity is generated by burning fossil fuels. We are told to use less energy — turn off the lights, turn down the

> thermostat, closed the refrigerator — personal practices. We don't choose to get electricity from burning fossil fuels when there are alternatives — the industry does. If we change at the industrial level, our personal choices matter far less.

This continued translation of bad business and industrial practices into an attack on personal practices happens for a reason — it takes the pressure off of business and industry. Here's the problem: ***that's exactly where the source of the problems is.*** If you cut down your consumption of meat by 10%, you might have a minuscule impact, but if agriculture were forced to stop deforestation altogether, it would have a huge impact. Where should you focus your energy? On your personal life? Or on getting agriculture to stop deforestation? If you stop your flight for your once-a-year vacation, you may be contributing, but if an airline changes to green hydrogen to power its planes, that is a big contribution. Where should you focus your energy--on self-denial or on helping airlines and other businesses make the transition? If you reduce your consumption of electricity by 20% by adjusting thermostats, you will make a contribution. If your electricity comes from renewable sources, however, the amount you consume matters far less from a climate change perspective. The carbon has already been eliminated.

Do you see? It is a disingenuous con job to get everyone navel-gazing at their personal activities when the world is being destroyed at the hands of forces much larger than we can address in our personal lives.

To get an idea of how people think about this, I asked readers on Medium to explain how reducing consumption can solve climate change. Here's a summary of what I discovered.

There are the three distinct schools of thought in the climate change writers' community. Roughly speaking, they are A) the group that wants to overhaul how people live and advocate the reduction of consumption by 50–70% mostly at the personal level; B) the group that sees the "consumption" in the system and advocates reducing it there; and C) the group that puts more faith in technological fixes like renewable energy and electrification. So you can recognize them, I offer these portraits.

### *Advocates of Personal Consumption Reduction*

Primary to the advocates of personal consumption reduction is the contention that consumption reduction is essential to solving climate change. As mentioned before, one guy went on and on about people idling their cars at fast food drive-ups. Another was incensed that people drive their cars at all. Yet another visualizes eliminating all private transport, developing regenerative agriculture, and eliminating meat as a dietary option. Another says: "Stop growth and go into low level maintenance. No new anything. Everything has to last. Reduce population, demographic decline. Dismantle old stuff when no longer needed in order to obtain new resources. Manual labor to replace much machinery. Work slowdown. Stop all make work…" You get the picture. Most folks in this category believe that people in the West, especially, can "reduce their consumption by easily 50% with minimal lifestyle impact…" Some say 70%.

I asked one such commenter how one would get home with six bags of groceries for the family if he didn't have a car. He sent me a photo of a backpack. I didn't follow up with a question about getting lumber for a home remodeling project.

What I notice in many of these comments is an animus toward modern life, as well as capitalism. "Capitalism drives consumption so we must abandon that system…" And later in response to someone else's comment: "But the top 10% must end their excessive lifestyles, committing billions to death."

A second thing I notice is a total lack of a plan. Most people's essential carbon impact comes from commuting, eating, and heating or cooling homes. The second ring of consumption lies in entertainment, travel, socializing, home décor, and home improvements. A third ring derives from their work — the heating and cooling of that building, the expenditure of emissions to produce goods or services, etc. The advocates of personal consumption reduction offer little in the way of what and how to reduce here, except for "drive less, fly less, eat less steak, outlaw advertising, and otherwise voluntarily spin down their lifestyles…" "Spinning down lifestyles" is not a plan. There are no objectives, no goals, and no milestones. There is no indicator as to how the individual has done given their initial baseline, and as we have seen, there is no way to measure if it is enough.

Although I have many thoughts as to why this is the case, suffice it to say that a position advocated without a plan on something as important as climate change is, at best, a red herring. I would challenge these advocates to articulate their real plan and consider the impacts. For example, when someone says we should just do away with cars, a million questions come up. What happens to the employees of car companies, auto repair shops, gas stations, and car rental companies just to get started? What happens to the millions of families whose most valuable asset — their home — is wiped out because of the collapse in suburban real estate prices? How will it be better that parents, who used to be home before their kids, now are stuck on trains and arrive home hours after their children get home from school instead of being there for them? I'm not saying there aren't answers, nor that they might even be very good answers. It's just that advocates of this position need to deal with the answers to be taken seriously.

I should also note that other commenters indicate that they have serious reservations about the consumption reduction programs and mandates as well. One said, with perhaps more attitude than was necessary, "Reorganize all of humanity in eight years to make you feel better. Got it. Stop growing enough food for everyone. Got it. Stop travel, and prevent people from earning a living. Got it. You really should check out North Korea." And another said, "I can see how those suggestions add up to move all the needles in the right direction… but those are BIG societal, psychological changes. I think we'd see fascism blow sky high in rebellion and burn it all down rather than have any of this "forced" on their choice to consume… People might want the results, but can they make the changes? Probably not."

In many ways, this little microcosm captures the essential problem with the pure consumption reduction position — it evokes such resistance that it is simply not practical as a solution. Any realistic solution needs everyone on board. It cannot be that there are do-gooders who sacrifice so that others can play. I get that this is already how things are, in a way, but as a solution to advocate, this will get us nowhere near the participation that is needed.

## *Advocates of Consumption Reduction at a Systemic Level*

There was a group of more interesting comments about consumption reduction at the systemic level. One effort to define such consumption was made by a commenter thus: "What climate scientists are talking about when they talk about consumption is a nation's material footprint: the total mass of material extraction within its borders, plus the mass of its imports, minus the mass of its exports." The discussion of consumption reduction as a strategy would help if this definition were widely used. It would require a systemic look at things. For example, I don't know how much oil, gas, and coal production adds to the equation for "total mass of extraction" and "mass of imports" but I imagine it is significant enough to radically change the calculation for most industrialized nations. Ostensibly, those numbers could change without the lifestyle changes the first group insists are needed anyway. Convert the entire automobile fleet to EVs, use electricity from non-fossil fuel sources, and the driving can remain while nation-level consumption decreases.

Several commenters saw that the notion of reducing consumption as a straight-line impact on emissions reduction was implausible. One primary reason cited was the growth in population in the developing world, which will easily outstrip any gains from conservation in the industrialized world. As one said, "movements like minimalism or such are not going to achieve the changes we are seeking." Another figured that 10% of the reduction in emissions we need may come from consumption reduction, but the other 66% would need to come from technology replacement. These folks were interested in seeing through to the causes, as well. By restricting choices upstream, you can change things downstream without much notice. "…if you go to the supermarket and you find cheap products wrapped in plastic, you are likely to buy them. But if you don't find any plastic, you won't even think about it."

## *Advocates for Technological Change*

Proponents of new technology are putting their confidence and hopes in technological advancement creating the change we need. A good example of this attitude is in this comment: "As the developing world modernizes, overall consumption will likely increase, not decrease — so "other" solutions like clean cement and clean steel will become more important. R&D to make those technologies affordable is key, as well as smart grids and other infrastructure to make renewables workable." Another said, "So, according to this one policy adviser, the answer to your question seems to be the fourth option, 0% from consumption reduction, 76% from renewable sources."

I am skeptical that we can reduce our way out of climate change. Humanity aspires to live what they see as a better life, and most of the time, that perception results in more consumption, not less. Since population is growing, any consumption-only strategy is doomed to failure. If we remain dependent on carbon, using less means emitting less carbon only until the population develops and catches up, and then we will be right back where we started as the additional people consume to meet their needs. A 10% reduction in consumption is replaced in a few years when the population is 10% higher, even if it is consuming at that reduced level.

Rather, we need to get the energy and power we need in a different way, and technology is critical for that. Electrification and renewable energy are leading the conversion, but deep in the roots of our industrial society, we have other processes — like cement and steel production, how we obtain minerals and materials, how we produce food, etc. — that will also make huge differences that are unseen to most of us, but critical for our world.

My conclusion after finishing this exercise is straightforward. In order to reduce carbon emissions by 76% as prescribed by IPCC using a consumption reduction strategy only, you would have to reduce consumption by 76%. Individuals will never be able to do that, nor will societies. Even if you could get three quarters of your people to do it, those who don't participate leave you 25% short of the goal. Advocates for pure consumption reduction have either not thought this through or have a different agenda they are not stating. Once again, we have a solution that is looking for a problem to claim as its own.

# Why "Being Scientific" Is a Losing Strategy for Climate Action

As a new student at the University of Minnesota, I read my first scientific article about climate change in 1980. It was published in 1979. I read it as part of a literature review I completed for a class I had in environmental geography. All the articles were published in the 1970s. At that time, scientists called it "the greenhouse effect" and discussed how it would lead to global warming. Only later did the term become climate change.

Since that time, the statistics have mostly gotten worse. CO2 emissions, for example, increased by over 90% since 1970. A recent report by the IPCC says: "Since 1970 the global average temperature has been rising at a rate of 1.7°C per century, compared to a long-term decline over the past 7,000 years at a baseline rate of 0.01°C per century (NOAA, 2016; Marcott et al., 2013)." It provides the data to back it up. EPA demonstrates the results of these changes that are observable in the world: "The planet's oceans and glaciers have also experienced changes — oceans are warming and becoming more acidic, ice caps are melting, and sea level is rising." And, of course, there are well-documented floods, droughts, or intense rain, as well as more frequent and severe heat waves that resulted from climate change.

The science is clear. It's been known since the 1970s. It was popularized more broadly with Al Gore's Inconvenient Truth tour in 2006. Yet here we are fifty years later, and sixteen years after Gore's publicity push, and everything is getting worse. Why?

There is a truism in sales and marketing, the area in which I spent most of my career. Here it is: "Facts tell, stories sell." I can think of no better example to demonstrate the reality of this truism than the climate movement of the last 50 years. We have been pounding people with facts. Countless academic studies, think tank reports, Al Gore's work, the Rocky Mountain Institute, The Sierra Club, and so many more. Facts, facts, and more facts. It's been good. The knowledge is sound. We know the science. And yet…

25% of Americans — i.e., 83 million people — either doubt or deny the science.

Read that again. After fifty years of well-established science…

25% of Americans — i.e., 83 million people — either doubt or deny the science.

It is safe to assume that those 83 million people will do nothing to alter their carbon-rich lifestyle. They won't stop driving, they won't stop flying, they won't reduce their meat consumption, they won't degrow, and they won't go solar or buy an EV because of climate change. They don't believe there is climate change, so how can it ever motivate behavior change?

One implication is obvious. Anyone who quotes science to say, "The science says that if we all did X we could reduce carbon by Y," must discount that carbon reduction by at least 25% because 25% of the population refuses to go along. In addition, depending on the action selected, many more may find that particular action unpalatable for a variety of reasons, and also choose not to go along. Could be an additional 1% or an additional 50%, but it will be more. So even if the science says that if we all do one thing to reduce carbon and it would have a big effect, human behavior is not going to comply. Too much of the population is not motivated to change, so the benefits claimed from such an approach will never materialize.

Science needs to continue to give us data and information — indeed, it is critical to do so. But as a weapon of *persuasion*, it had its chance. Fifty years of chance. Despite that, emissions increase, consumption increases, global temperature increases, etc. So what do we need to do if asserting science can't achieve all we would like it to achieve?

We have two options: Expand the motivations for the desired consumer behavior or change the system to eliminate consumer choice.

## *Change Motivations*

Using science, we can determine the behaviors we want to see adopted across society. We can make the case for those things scientifically to those for whom that method is effective. But for those not persuaded by the scientific argument, we could generate new and different motivations. For example, we know that electricity generated from solar arrays does not create emissions, so we could determine that it would be good for everyone to have solar panels on their home. Some people motivated by the science of climate change want to do their part and adopt solar panels on their rooftops. That's pretty much how it is today — some people have already done this. Why haven't more people done it? Well, 25% don't believe the climate science, so they have no motivation whatsoever. Many more don't have the upfront cash to put into it, so they hold back. More still are risk-averse regarding their roofs and their homes. What if they need a new roof? Moving the solar panels to build the roof and putting them back could be very expensive. Plus, people feel that if they make the investment and others don't, there might be no impact on climate change anyway, so in their view, they just end up with an expensive inconvenience and climate change goes unabated.

An autocratic government like China's could certainly force everyone to install solar panels, but that is unlikely to ever happen in the US. The American alternative, however, might look at all these objections people have and find a way to mitigate them. Instead of tax credits for solar, which tend to favor people with enough income to benefit from tax credits, direct subsidies could be provided. Financing could also be provided in such a way that it guarantees a person's electric bill will always be significantly lower than it would have otherwise been. Perhaps government insurance could be provided regarding roof replacements and the removal and replacement of the solar array. These ideas illustrate a way of changing people's motivations. When putting solar on your roof goes from a large investment requiring owing taxes to get the tax credit and high risk (in the form of roof replacement or changes in electricity costs) to an obvious financial win with no risk and lower costs, people will do it. It doesn't matter what you believe about the science any more —

you'll do it because it makes financial sense. The motivation has changed.

Any carbon reduction strategy recommended by the science will see increased adoption when viewed through the lens of human motivations. One's judgment about those motivations is not relevant. What matters is how the person you want to change their behavior views the problem. Wherever we can create non-science-based motivations for climate-enhancing behaviors, we greatly expand the number of people willing to participate.

### *Eliminate Consumer Choice*

The other option is to eliminate the anti-climate choice altogether. Today, for example, people can choose to buy a traditional internal combustion engine (ICE) car or an electric vehicle (EV). Likewise, many utilities provide power from traditional sources and they provide an opportunity to subscribe to a community solar garden or wind farm as an alternative. In both cases, the same 25% who don't believe the science is unlikely to adopt the new climate-friendly product. Why should they? The non-climate choice has been effective, certain, understandable, and in most cases, less expensive. Those who do believe the science may or may not adopt the new products based on other motivating factors. The good thing is that people have a choice. The bad thing is also that people have a choice.

It wasn't always this way. Both the automobile industry and the electric generation industry have offered one and only one alternative for decades—ICE cars and electricity from fossil fuels/nuclear. Did you want a car? Internal combustion was the only option. Do you want electricity? You get what they give you by burning fossil fuels or nuclear. There were no options.

While it is good that we now have options, it is better that we are moving beyond options. We should accelerate this trend. There is no reason why the one and only one choice can't be the climate-friendly option. What if there are no ICEs available at all, and the only option for a car is an EV? And what if the only electricity you can purchase comes from renewable sources because no one burns fossil fuels anymore? The good news is that this is happening.

Car manufacturers are making this change themselves. Audi says it will offer only EVs by 2026. Honda will only produce EVs by 2030. GM is making its Buick and Cadillac brands 100% EV by 2030, and all others by 2035. In effect, the industry is moving to eliminate the climate-destroying option because of perceived market forces. But the effect is that consumers will not have the choice of buying gas-burning vehicles.

Renewable electricity generation is lagging far behind where the auto industry is taking us. Wind and solar are growing rapidly, but they still account for a small fraction of global electricity generation. However, as the mix changes and renewables become 20%, 50%, and 80% of the electricity source, consumers buying that electricity no longer have a choice of buying the polluting, high-carbon variety. And they won't care.

The new cars and the electricity generated from renewables will give people exactly what they expect as they have come to expect it. Cars get them where they are going reliably and safely, and electricity turns on the lights when they need it. But here's the difference: That 25% who don't accept the science? They are buying the solution anyway. That 1% to 50% who might otherwise not choose the climate-friendly alternative? They are buying the solution anyway. Participation in the solution becomes 100% because participation is involuntary. There's no coercion. It's just that there is no other choice.

### *A More Holistic Approach*

Science has its role. We need science to define the problem and tell us what will work. Using its data and insights, we need to determine where the biggest impacts can be. People need to understand what will work best. But to get people to adopt the solutions, we need to turn to motivation, marketing, and the study of human behavior. Part of this is to restructure the incentives, part of it is to bake the solutions into the system, and part of it is to repackage the already existing facts. That's called marketing. It is absolutely essential to changing the human behavior necessary to solve climate change.

# Changing America May Not Be the Climate Solution We Think It Is

The climate change debate is fraught with antipathy toward Americans. We are greedy, oblivious, and hateful. We are gluttonous, fat, overfed, and wasteful. We are self-centered and don't give a damn about anyone else. Generally, the attack takes the form of what a commenter said on a Medium post:

"I really hope I'm wrong, but does anyone really believe that Americans (generally perceived as the most self-centered and oblivious people on the planet) would be willing to drive less, fly less, eat less steak, outlaw advertising, and otherwise voluntarily spin down their lifestyles to save the planet from a vague future threat that many of them are being told is fake news anyway? Remember, these are the same people who were willing to let a million of their fellow citizens die to avoid the inconvenience of wearing a face mask. And who among the current pantheon of American political figures would have the courage to propose such a path?"

I'm an American, and to be honest, I don't take exception to many of these characterizations of us as a society. I understand, I think, how it looks to others who are concerned about climate. My country led the way for 250 years of burning fossil fuels. Today, we emit far more carbon per capita than any other nation. America created incredible wealth by not counting a bunch of costs we are now seeing as climate change. We insisted on globalization and expansion to continue the growth path capitalism requires. I get it. I really get it.

But as well as being an American by the simple luck of birth location, I am also a citizen of the world, and I am deeply concerned about climate change. I know we need to do the right thing, and I want to do the right thing. Most people I know want to do the right thing.

People implore us to strike out individually to make the necessary changes. Change your lifestyle. Go local. Stop flying. One commenter, for example, put all the blame on cars — a not-so-subtle dig on how most Americans live in and for their cars. I asked,

"Out of curiosity, if all cars are eliminated, how do people get around?"

The response…

"Walk. In large cities, busses and trams. Between settlements, trains or busses. If that doesn't seem feasible, then it's because you're envisioning a society set up to cater to cars, not people."

Pardon me, but it is not at all that I am "envisioning a society set up to cater to cars," but rather that American society is already set up that way. It was set up when I was born into it, and it has been set up that way all my life. Of course, there are alternatives. Of course, societies can be built and structured in other ways. But ours is not. We have inherited and added to an infrastructure of tangible reality and cultural reality built on cars, to be sure, but even moreso, on fossil fuels. We have skills, habits, and assumptions about how our world works, and most Americans have mastered that to one extent or another. We are structured to live that way, and not any other way.

When the climate change debate implores Americans to "voluntarily spin down their lifestyles to save the planet from a vague future threat" we are moving into the area of impossibility. Individual choices to work against an enormous system are laudable, but it is not practical as a way of getting to the change we need fast enough. Individual change in this environment is a huge lift with an extremely small impact. Go ahead. Sell your car. Walk everywhere. Buy your backpack for bringing home your groceries. Avoid the events, classes, and meetings that make life interesting that are on the other side of town because there just isn't enough time to get there by bus. (Skip your anti-climate change activist meeting, too.) Take your kids out of sports, music, art, and church activities because you can't get them there. Do the big lift and make the big sacrifice. How much impact will your effort result in? You shrink the carbon footprint of one family out of the 8 billion people on Earth. It is a pimple on an elephant's ass, at best.

### *A Better American Contribution?*

Let's keep in mind another feature of Americans — we are pretty damned ingenious. For all of our faults, Americans are creative. Americans are good at creating solutions and manifesting them in the world. We call it entrepreneurship. What if instead of attacking Americans to reduce our lives in ways we do not know how to do with minimal impact, we call on Americans to get creative, put our minds to solutions, and change the world the way we know how to do — through business, technological advancement, and creative entrepreneurship.

I know many people in the climate debate seek system change. Capitalism is evil, we consume too much (especially Americans), the world produces way too much junk, and growth must be sacrificed. In all honesty, a more pedestrian society would be quite attractive to many people. But all that system change is an enormous undertaking that would require decades of planning and building, trillions of dollars in public investment, and probably a ton of litigation. Transformations like this involve winners and losers, and you can bet that the losers will resist. Just imagine, for a moment, what would happen if the car hater quoted above got his way and all cars were taken off the roads. Can the bus system in your city accommodate hundreds of thousands, or even millions, of new riders? How about the trains? How many new buses would need to be built? Where would they come from? How fast could they be built? And how much carbon would be emitted into the atmosphere in the effort?

And then there are the social-economic considerations: Unemployment in the auto industry, wasted time on longer rides and waiting for buses, and the sudden uselessness of displaced suburban homes. Can you imagine this might have political repercussions?

Here is the worst part: We could go through all this turmoil and disruption, totally change our lives, restructure society, and dislodge capitalism as a system… **and it would not stop climate change**. Why? Because that kind of system change can't happen fast enough. It is physically impossible to change a society's infrastructure that fast. Many would say that cultural change is even slower. We just cannot get there this way.

The antipathy toward Americans, whether earned or not, doesn't help us get to a solution. Rather than imploring Americans to shrink their lives, let's motivate them to commit everything to solve this challenge. Let's urge all Americans to give up their jobs producing stuff people don't need, and commit their careers to jobs that will solve climate change. Let's put this discussion on the lips of everyone in every place in the American and global economy. Let's utilize the structure we have to make the solutions we need as quickly as possible, so as to provide the opportunity for the longer-term system change to be decided and manifested. Let's make solving climate change honorable, worthy, and sexy. We must unleash the creativity of Americans and the American system to fix this thing. We can build out a 100% renewable energy system. We can figure out how to capture and sequester carbon, or better yet, use it. We can build a system that will use less of everything, thereby enabling everyone to use less whether they believe in the effort or not. This is how we can get to a solution.

# Billionaires Are Not the Problem

While most people continue to frame the climate debate in oppositional terms—i.e., people against oil companies, business, and capitalism—we've just seen how the interests of climate activists and those of these same business entities are aligning. This alignment forces thoughtful people out of their ideological positions and into real engagement. Nonetheless, there seems to always be a human desire to blame others. Next on the list? The billionaires. A "study" by OxFam put them right in the crosshairs, and I will say it now: This study is an outrageously stupid distraction.

Here's the headline on CNN:

Billionaires emit a million times more greenhouse gases than the average person, study finds

What do you think immediately? The wealthy are to blame for climate change.

According to CNN:

"Nafkote Dabi, climate change lead at Oxfam, said that emissions from billionaires' lifestyles, their private jets and yachts are already thousands of times higher than those of the average person. But looking at emissions from their investments, their carbon emissions are more than a million times higher."

This information is a distraction. It implies that if we could just change the billionaires, we could solve the climate problem. It offers a panacea where none exists. And, conveniently, it let's everyone else off the hook. "It's not my fault! Let's just fix *those* bad people!" A nifty psychological trick, indeed.

Did anyone stop to think that a person with a billion dollars of net worth is going to have a ton of investments in operating companies? Did it occur to anyone that in today's world, those companies, like all companies, generate emissions? I mean nearly all economic activity that uses energy today results in emissions, and it doesn't really matter who owns it. If the activity goes on, we'll have emissions. It doesn't matter if the company making that activity is owned by a billionaire, its employees, the government, or a non-profit. The *activity* is the problem.

Here's the thing. If the billionaires divested ALL of their investments and sold them to mutual funds in which billions of common people invest, **there would be absolutely no reduction in emissions whatsoever**. The companies would keep operating, just exactly as they are today. *Total emissions would not change.*

For that reason, studies like this are an abomination. They distract us from the real issues that will make a real difference. One can imagine this study becoming a topic of conversation at a COP meeting, for example, where much bigger and more important issues should be attracting people's attention. It won't be hard to imagine activists going nuts over the contribution of billionaires to climate change. They'll read the headline, create a meme, and soon that becomes the focus. Yet if action were taken to force billionaires to sell these investments, the net change would be zero.

The distraction is hurting the cause. When so much necessary work is still waiting for attention, people working on climate change do not need this kind of distraction. We need to be researching technology, deploying solutions, and changing the world's energy system. Demonizing billionaires does not help.

"Studies" like this become a lightning rod for stupidity. Watch the anti-climate media go nuts. This is fodder for their conspiracy theories, denigration of the climate movement, and long interviews that lead people to believe that the entire climate movement is a façade for a socialist takeover of the world.

This is decidedly unhelpful. While inequality is a societal challenge that needs to be addressed, the demonization of the wealthy moves nothing forward except envy, anger, and greed. Those of us who are not billionaires start to blame those who are, but that is not a solution. It fixes nothing. It doesn't convert energy to renewable sources. It doesn't capture carbon. It doesn't change the livestock industry. It doesn't help the adoption of electric vehicles that can be powered by renewable electricity. And what is the net result? A feeling of powerlessness, angst, and resignation. Animosity rises. It becomes harder to work together.

So if you are an activist, don't get sucked in by this. We need you. We need your energy going in a helpful direction. Whatever role you are playing, accelerate it. Whatever commitment you have made, deepen it. As far as the wealthy are concerned, it might be far better to connect to the wealthy to help them invest with the rest of humanity rather than demonizing them. It might be better to work with everyone we can work with rather than demonize some, especially those with the means to make a difference. I have no idea what Oxfam is trying to do with studies like this, but for people of goodwill who want to solve climate change, this is a step in the wrong direction.

# How Earth Will Solve Climate Change If We Don't

If you have not read the opening to *Abundance* by Ezra Klein and Derek Thompson, you should. They paint a picture of exactly how human and American ingenuity ought to bring about a new world that is largely a solution to climate change. It is a beautiful picture of society running on alternative sources of energy, pollution free, in a dynamic economy creating wealth, housing, and well being for everyone. Beautiful.

On the other hand, if you have not read *The Ministry for the Future* by Kim Stanley Robinson, you are missing another compelling opening, this one illustrating a fictional but colossal disaster in a heat wave in India that kills twenty million people. It shows the utter break down in society, the break down in the electric grid, and the impossibly haunting image of thousands of people floating on a lake trying to cool themselves—except that they are all dead.

These two openings could not be more divergent. My book about climate change, which I collected together from two years of writing in 2021-2023 argues from its core that we can solve climate change with technology, and that in fact, we have no other choice. Klein and Thompson have shown a positive view of what is possible. They outline the vision I want to believe in. But as the changing climate has unfolded, and as we have seen efforts to curtail carbon emissions fail to meaningfully reduce them globally, and as we have seen successful programs and investments retrenched, any neutral observer must conclude that the scenario in Robinson's book is much more likely than that in Klein's. In fact, what's likely is far worse.

In 2024, global GHG emissions increased again to the highest level ever. Atmospheric carbon was 280 ppm in 1750 as the dawn of the industrial revolution when we started burning fossil fuels. Today, it is at 425 ppm. Pretty much everyone knows that's because of GHG emissions from burning fossil fuels. It seems that the solution is to stop burning those fuels, yet from a practical standpoint, it is impossible to simply stop without a replacement. People would freeze or roast in their homes, offices, and factories. Or they would starve because store shelves were empty. Or they would die of thirst because water systems can't operate. As a species, we burn fossil fuels because that is the only way we can continue to survive. To stop without a replacement would produce a massive reduction in population—clean words for mass death.

The thing is, this solution may not be so wrong. Mass death may, in fact, be the only way to finally reduce GHGs. About nine months ago I did the math for another article I published, and what I found was stunning to me. I use the year 1750 as the start date—the beginning of the industrial revolution. Here is how I walked through it:

The global population estimate for 1750 is 750 million. Today, it is 8.03 billion. That is an increase of 10.7 times from 1750 to present day. During that same time, estimated global CO2 emissions went from 9.1 million tons in 1750, to 37.1 billion tons today. That is an increase of 4,078 times since the start of the Industrial Revolution.

At first, I thought the 10X growth in population was dwarfed by the overall increase of 4,078X. But then I did the math.

Global emissions since the Industrial Revolution grew by 4,078 times. During that time, population grew by 10.7 times. If you divide out population growth (4,078/10.7) the factor left is 381. That factor represents the growth in emissions from influences other than population growth. Multiply the carbon emissions in 1750 (9.1 million tons) by that factor of 381, and you get 3.468 billion tons of CO2 emitted into the air. That seems like a lot. But today, our current global emissions are 37.1 billion tons! The difference? That 10.7X growth in human population.

One could guess reasonably enough that the 381X factor has to do with the expanding adoption of fossil fuel technology—i.e., more and more people driving cars and electrifying their homes—and the development of new technologies that require ever more energy—for example, cars replacing horses over a century ago and the more recent developments of AI and cryptocurrency. Perhaps there are other factors, but these two are a big part of the 381X factor. Despite that, without the increasing number of people on the planet to adopt and use fossil fuel driven technology, our emissions would be 90% less than they are today. We would not have climate change.

In addition, there would be no UN IPCC. Indeed, there would be no problem.

Furthermore, if population had remained stable and every one of the 750 million people or so on earth was now living a life similar to the US lifestyle—that is, with maximum use of fossil fuels per capita—total emissions would be about 75% less than they are today. Again, there would be no problem with climate change.

Based on this analysis, one can see that the two book openings I cited at the beginning of this essay actually reflect the potentialities we face. We either achieve a massive change over of our energy technology to be emissions free, or we face the likely scenario of population reduction. Let's look at both possibilities.

### *An Energy Technology Makeover*

To survive at the population scale human beings have achieved, we must have energy. It not only heats and cools our homes, factories, schools, and offices, but it also grows our food, delivers our food, and cooks our food. Energy gets us the water we need and the materials with which we build. It transforms materials into the things we use every day, from vehicles to computers to shirts and shoes. Indeed, the massive growth in our population could never have been achieved without the access we have enjoyed to cheap, abundant energy. To pretend that today we could just do without it, leave the oil in the ground, or go back to simpler times is only to fool ourselves into accepting a catastrophe.

To maintain our population level on the planet we need to change the form in which we obtain and use energy. Human beings already have the technological capability to make this change. It requires that we move away from burning fossil fuels and toward three other sources—renewables such as wind, solar, hydro, and tidal; alternative fuels like hydrogen; or nuclear. Efforts have been made in these directions and some success has been achieved. Today, for example, solar and wind powered electricity are the cheapest forms of electricity worldwide.

The problem is that most of our societal infrastructure is not currently designed to be electric. From automobiles to home heating to military defense weaponry and industrial and agricultural systems, we burn stuff to live. It's not just that we need to create the electricity in new ways—it is also that we must transform everything we do in society so that it can operate with clean electric energy or clean hydrogen fuel. That means radical changes in the economy. Government, churches, schools, businesses, homeowners… all of them are invested in the burning technologies and all of them would need to transform. Not just in the US, but everywhere.

The challenge, however, is that as we have delayed for decades, the climate has already started changing, and those changes are beginning to make things irreversible. Coral reefs are dying off, which means that fisheries are not nearly as productive as they once were. The boreal forests of the north are burning, and as they do, the land is exposed and the greenhouse gases locked in the soil are being released. The Amazon rain forest, sometimes called the lungs of the planet, is shrinking so that it has lower and lower capacity to convert carbon dioxide to oxygen. Hence, even if we convert everything today and completely stop emitting GHG into the air, newly enabled sources will continue to emit. And what we already have in the air will continue to trap heat and warm the planet until even larger impacts, like the slowing and stopping of the Atlantic ocean current, known as Atlantic meridional overturning circulation (AMOC), occurs.

We are at a point, in other words, where ending emissions is not enough. We need to figure out how to get what is already in the air out of it. This is known as carbon capture technology or carbon sequestration. Humanity has the technology to do this in conditions of concentrated air flow, such as smokestacks and exhaust chimneys, but the technology to do it at scale in the atmosphere is limited and extremely expensive. It is hard to see that we will ever figure this out or take the right actions. No wonder men like Elon Musk want to escape to Mars. There seem to be no solutions we can develop and deploy in a short enough time frame to avoid the catastrophe we have made for ourselves.

*The Population Reduction Solution*

If we cannot lower the level of greenhouse gases in the atmosphere, we cannot stop the heating. Climate may change in erratic ways, but the steady march toward increased heat is already creating events that are not survivable by human beings without artificial cooling—i.e., air conditioning. There comes a point in the combination of temperature and humidity where the human body can no longer cool itself. When that happens, core body temperature just keeps increasing and it eventually sickens and kills the person. This is the scenario at the center of Robinson's book opening.

There are various ways of understanding this level of heat. There is the regular temperature, known as dry bulb temperature. There is the heat index, which tries to describe show hot the air will feel. And there is the wet-bulb temperature, which measures the body's ability to cool itself under the conditions mixing temperature and humidity. The wet-bulb temperature is what scientists us to measure survivability of heat waves. The maximum survivable wet bulb temperature is 35 °C (95 °F), so when we get to that level, we have gone beyond survivability.

The thing is, it has already happened, although briefly, and more and more instances are occurring in which humans cannot survive the heat. Here is a summary:

- **Jacobabad, Pakistan & Ras al Khaimah, UAE** reached **~35°C wet bulb** temperature for short periods in 2021. These were the first verified instances on record of having surpassed wet bulb survivability.
- **India and Pakistan,** various heat waves from 2015–2023, brought the wet bulb temperature **to 33–34°C**.
- **Persian Gulf coast & Red Sea**—coastal cities like **Doha, Dhahran, and Bandar Abbas** regularly experience **>33°C wet bulb** during summer.
- **Southeast Asia (2023, 2024):** Heat waves in **Thailand, Laos, Vietnam**, and **Bangladesh** recorded **wet bulb temperatures over 33°C**

For Americans, these conditions can feel very far away. And yet, they are not. We know that hot weather has been increasingly common in places like Phoenix, but in 2021 a heat wave hit the pacific northwest and temperatures in Portland hit 116 °F. The actual wet bulb temperature got to 90 °F (33 °C). Lytton, BC in Canada surpassed 121 °F. These are areas where AC is not prevalent because it is generally not needed. Depending on the data cited, at least 800 people died unnecessarily. Some estimates put it at nearly double that.

If we do not solve climate change—i.e., reduce atmospheric carbon radically and in very short order—we are going to create more and more of these kinds of events. And in that future lies a catastrophic solution to climate change—mass reduction in the human population.

There are two ways this catastrophe could unfold. The first is that it just gets so hot that anyone exposed to the heat will not be able to survive. A heat wave like this doesn't have to last that long to kill a lot of people. It might seem that those with air conditioning are safe, but really they are not. Air conditioners, like utility pole transformers, are engineered to work well at certain levels of ambient air temperature. My air conditioner is rated for 108 °F. Most utility pole transformers are rated for 110 °F. The heat waves we are talking about are pushing ambient air temperatures to over 120 °F and higher. It doesn't take a genius to figure out that at some point, the AC doesn't work anymore. Either your unit breaks or the grid breaks, but there is no longer a way to cool yourself. In these conditions, without cooling, human beings overheat and die. Imagine a Phoenix heat wave in which the electric grid collapses. No one has AC. Temperatures soar to 135 °F, and at night it only cools to 110 °F. Without AC, no one survives.

The second way this could unfold is through the heat's impact on agriculture. Humans are not the only mammals that cannot survive high wet bulb temperatures. In fact, the survival threshold for our most common livestock—cattle, chickens, pigs, and sheep—is much lower than it is for humans. While humans can survive up to 35 °C wet bulb temperature, the limit for livestock is mostly between 20-25 °C wet bulb temperatures. Likewise, our crops are even lower than that—especially the cool weather vegetables like broccoli and lettuce. In other words, as the average temperatures increase, our ability to feed humanity will decrease. Overall agricultural production will shrink. Even if you are not in a heat wave, food shortages will affect you. As a species, we will begin to starve as all species have throughout time. We will overeat what food is available and leave many behind to die.

## *Scale*

One way or another, GHG emissions will be reduced eventually. Either human beings will do it with technology invention and adoption or nature will do it by reducing the population on the planet until emissions are reduced. Since humanity appears to be unable to stop emitting GHG into the atmosphere, the earth's systems seem poised to balance itself by stopping humanity in its tracks. All of its systems seem to be working toward that balance, and central to it seems to be a reduction in life, especially human life. Population will be reduce. Heat waves, I believe, will do more to reduce population and restore balance than any other change. But storms, droughts, floods, and fires will play their parts as well.

The questions in my mind is *how much must the population be reduced?* We are now at 425 ppm CO2 in the atmosphere. Scientists say we need to be at 350 ppm. The UN IPCC says we would need to reduce emissions by 75% by 2030 to get to that target and avoid some of the worst climate change impacts. If everything else stays the same, that implies a 75% reduction in global population to get there. Catastrophic indeed. Mother earth would need to be very angry.

The reality is that neither of these scenarios is going to play out exactly. Humanity will continue to develop and deploy technological solutions. It won't be fast enough to eliminate all of the damage, but the faster we do so, the less violent the Earth's reaction will need to be. At the same time, the 2030 target is a human target, not that of the earth. Earth doesn't really care about the timeline. As far as we can tell, however, Earth's systems are tuned to enable it to survive. They are already responding to the conditions we are creating, and they will continue to do so on earth's timeline. There is unlikely to be an event that will eliminate 6 billion people quickly. On the other hand, the slow incremental squeeze against a human endeavor that cannot be sustained by the Earth is obviously inevitable. We are already in it. The slow shrinking of arctic sea ice. The slow melting of glaciers. The slow whitening of coral reefs. The slow increase in fires, storms, and droughts, and the slow increase in intensity of all three. The slow increase in the size, duration, and intensity of heat waves. The slow decrease in agricultural production. All of this will affect the human ability to live. Fewer and fewer will be able to survive, and that will be the slow unfolding of the catastrophe of the human endeavor on earth.

My God. I wish I never had to write these words.

# Is It Too Late or Not for Climate Action? The answer may surprise you

Even with business and capitalism strengthening the fight for climate solutions, and with more clear focus on carbon, many people still wonder if it is just too late. So today's climate discussion gets be caught between two critical points of view. On one hand, people are saying that Earth and humanity are already doomed. We can't turn the tide, temperatures have already risen too far, and the feedback loop cannot be stopped. The inevitable outcome is a climate catastrophe.

On the other hand, people write that we can still avoid the worst effects of climate change. They argue that meaningful action can still have an impact and we should be working toward it as quickly as possible. The underlying notion is that we don't have to face climate Armageddon the way the doomsayers say we do.

So which is right?

Anyone following the global climate situation knows that we are facing increased climate change effects. Those effects include dried-up rivers (i.e., Colorado, Mississippi, Po, Rhine, Yangtze, and others), wildfires (Canada, California, Europe), cities running out of water (Nelson Mandela Bay, Rio Verde Hills), floods (Kentucky, Pakistan), and heat waves (British Columbia, Europe, US, China, South Asia). Agricultural production — that is, our food supply — is under pressure with examples like Spain's olive harvest down 40% this year, US corn production is down 8% from the previous year, and fears that other commodities are also down. Some areas, like eastern Africa, have drought so severe it is leading to mass migration and starvation. The Thwaites Glacier, also known as the doomsday glacier, is hanging by a thread.

Even though all these things are happening, most people hardly know the difference. We hear the stories, but our day-to-day lives go on unaffected directly. We don't face drought, we don't face fires, and we don't face hurricanes. No one we know is starving or running out of water due to climate change. So it comes to us as the terrible news always does — as someone else's problem. War, injustice, crime, hunger, social disaster… all the news happens somewhere else. While we are horrified by what so many people must face, we do not face it ourselves.

In other words, whether or not it is too late to deal with climate change depends on two key questions: Where do you live? And too late for what?

If you were flooded out in Kentucky or Pakistan in 2022, or in Vermont in 2023, it is already too late. Your home, your belongings, and maybe even your family are all gone. In that context, what does it even mean to ask the question: Is it too late? Climate change already destroyed your life. If you live in a zone that has already undergone a climate disaster, it is already too late to avoid its worst effects.

If you lived in Paradise, California in 2018 when fire swept through, killed 56 people, and destroyed the entire town, it is already too late. If you lived in Lytton, British Columbia in 2021 and were one of the 600 people who died due to the heat wave, it is already too late. If you live in Lahaina, Maui, it is already too late. If you are one of the million-plus people in Somalia who had to move because of drought, it is already too late.

On the other hand, the implications of climate change could become even worse. The collapse in the snow crab population from 8 billion to 1 billion in three short years occurred because of a sustained, few degrees of temperature change in their environment in the Bering Sea. Snow crabs need their water temperature to be within 2–3 degrees of freezing. The Bering Sea has become warmer than that. The population declined by 87.5%. We have 8 billion people on earth, too. We are looking at a few degrees of temperature change in our environment. If such a change were sustained, could the human population collapse in a similar way? Most likely, it could.

Although they rarely say it, I assume that such an outcome is what people mean when they claim we can avoid "the worst effects of climate change." Indeed, an 87.5% collapse in the human population would be catastrophic. The most likely change that would create such an effect would be sustained changes in the wet bulb temperature (or heat index) or sustained heat waves like the one in British Columbia in 2021 that killed 600 or that one in Europe in 2022, which is estimated to have killed 26,000 people. Overheated human bodies die. In terms of the universe, we live within a very narrow band of temperature possibilities. Our core temperature needs to stay around 98.6 degrees Fahrenheit — certainly no more than +/- a few degrees. Too high of a wet bulb temperature, which is an outcome of climate change, makes it impossible to sustain the proper body temperature.

We need to act and we need to act now.

The only way we can do this is to stop burning stuff — oil, gasoline, diesel fuel, propane, natural gas, wood, or biomass. The cause of climate change is burning stuff. Solar, wind, geothermal, nuclear, and hydropower are our only alternatives for electricity. We also have geothermal for heating and cooling.

Each of us needs to do all we possibly can with the situation we have, but the key is this: **Systems must change!** We need to make everything run on electricity, and that electricity must come from non-burning sources. Every electric company must invest in these sources now! Every city needs to plan to replace natural gas with geothermal heating and cooling. The entire auto and truck industry has to convert to electric vehicles, which it is doing, but too slowly. We need to bring an end to all smokestack industry and turn it all into electric — including steel, aluminum, oil refining, chemicals, and food processing. I'll explain details later in the book, but there are many systemic changes we can achieve now.

So, is it too late for climate action? The answer is yes and no. For an increasing number of people in an increasing number of places, it is far too late for climate action. They have already faced the worst they can imagine. But on the scale of mass disaster and massive collapse of the human and animal populations, it is not too late. Action now can still make a big difference, and we should continue such action until we mitigate it or the worst occurs.

# Why the Climate Debate Keeps Getting Stymied

Let me begin with a little story I heard from the author John Hassler. He is an author of fiction books. One time, his country home had a backup in his sewer and his basement filled up with sewage about a foot deep. He called a plumber who came immediately. The plumber put on his rubber hip boots and headed into the basement to try to find out what the problem must be. After a while of wading around in the basement sewage, he called Hassler over to the stairwell to discuss options. He told the author that he would likely need a new septic system. The plumber asked Hassler what he did for a living. "I'm a writer." The plumber said, "So you write all day?" "Yes." Then the plumber, standing in the sewage a foot deep, looked up, shook his head and said, "God, I'd hate to do that for a living."

Different values — this is why the climate debate is so difficult. We all have different values and things that are important to us. One person would rather walk around in rubber boots in a foot of backed-up sewage, and the other would rather write. Neither wants to do what the other does. Some people want to be self-sufficient and do everything themselves, while others hire experts because they'd rather not do those things. Some people collect books while others want to travel the world. The variety of human experiences, desires, wants, and needs is endless. We can all see things in society that we think are frivolous, and what we think is frivolous is almost always the needs and wants of **other people**. We rarely see our own needs and wants as frivolous because we value our own interests. It is an easy cop-out to devalue the interests, wants, and needs of others.

### *The losing strategy*

The debate has therefore pitted us against one another, especially using the construct of carbon footprint. Carbon footprint perpetuates the fallacy that it is about your values vs. my values, and it prevents unified actions. Instead, one person says I am the problem, and another says she is the problem. Divide and conquer. We define the problem, all too often, according to our own values.

Hence, vegans often tell me that if we don't all become vegans, we cannot solve climate change. Degrowth advocates tell me that if I don't become self-sufficient and consume a lot less, we cannot solve climate change. As a solar advocate, I might respond that if you don't invest in solar on your home, we cannot solve climate change.

But do you see how each of these statements is coming from one's own set of values? It's actually an attempt to inculcate the other into taking actions according to your values.

**People! This is a losing strategy!**

The imposition of one person's values on another person with different values never goes well. It is experienced as a tyranny, and as such, it is rejected. Often, it is rejected even if the two people have some affinity over the particular value. That's because the reaction to the tyranny outstrips the agreement on the value.

If we could produce cars with zero emissions and drive them with zero emissions, is car use and consumption still the issue as far as climate is concerned? If we can produce beef without deforestation and the emissions of feedlots, is veganism — i.e. our choice of what to consume — still the answer? If our electric energy could be produced totally with renewables, is reducing the consumption of that electricity still a climate change solution? I mean, if carbon emissions cause climate change, and we can do what we want to do without producing carbon emissions, why do advocates still tell others to consume less? It's because consuming less is important to their values, regardless of its value to a climate change solution.

As we go around telling everyone else how to live their lives, we make enemies, not friends. In many ways, this has been the fuel of the climate denial fire. In right-wing media and in personal conversation, you can hear formulations like this: "They want everyone to go back to the land, and why? There isn't any climate change anyway!" Many climate change activists will decry the stupidity of these people, but the issue isn't their stupidity — it is the attack on their values. They reject it out of hand and find anything they need to defend what's important to them.

This is a human reaction. The insistence that you must change because the things you value in life are somehow wrong will always meet with resistance. We all find ways to write off or defend against things that offend us. And we always will.

We need to find a way to discuss and debate climate solutions that do not devolve into attacks on one another's values, needs, wants, and desires. Many activists are fed up and impatient. They see lawns as frivolous, while others embrace a lawn as beautiful. They see travel as frivolous (unless they are the ones flying to a protest) and they prescribe that others should curtail their travel, even though many of those others live for the opportunity to explore the world. Some say bookstores are stupid and ask why people can't re-use at the library, while others tell everyone they should ride bikes instead of driving cars — like in Cuba.

This is never going to work. All it does is leave us at each other's throats all the time.

### *How to find success in the debate*

We will find ourselves much more successful by advocating better ways of achieving the outcomes people want than we will by arguing to deny people what they want. Rather than telling people they should become vegans, let's change the way beef is produced. Rather than telling people, they should not drive, let's advocate the benefits of EVs. Rather than eliminating electricity, let's make it without carbon emissions by using solar, wind, and hydro sources.

In other words, the real problem to solve is the way in which the variety of needs, desires, and wants of people are met. Today, most are met with carbon-intensive methods. Tomorrow, they can be met with non-carbon emission methods of production. That is the ultimate solution — providing life in a better way rather than denying life opportunities to people.

Let's look for that common ground as much as possible. Let's find those places where we can meet everyone's needs in a carbon emissions-free way. Don't worry. If you like conflict, there will still be plenty of it. But the solutions are available, and we can help heal our cultural debate if we can find that common ground.

SECTION 2

# *What We Need to Do*

At some point, humanity is going to stop climate change. We don't know when, nor what we will suffer before we do, nor necessarily how it will happen. We don't know the exact technologies or solutions, and we don't know what cultural mores will change as we solve it. A few big catastrophes may push us finally into action, or we may go the way of the Alaskan Snow Crab. If the human population shrunk by 87.5% like the snow crab population did, we wouldn't need any technology to solve climate change. The population reduction alone would do it, but we would experience the very worst effects of climate change.

The alternative to population collapse, as I have suggested frequently in this book, is technological. Human beings need food, water, heat, cooling, clothing, and more. We met those needs with cheap energy from hydrocarbons for the last 280 years or so, and the result has been an explosion in population and the onset of climate change. Hence, the only way forward, short of a catastrophic decline in population, is a new way to meet the needs of humanity. We know we can't keep burning hydrocarbons, so we have to find another way. That's why technology is so critical.

Let's imagine it for a moment. We convert the entire US auto fleet to EVs. Cars become quiet, fun, less expensive to drive, and clean. The smog in cities gets cut by 75%, thereby reducing lung cancers, emphysema, and other diseases linked to smog. Coal and gas-fired power plants close down and are replaced by solar and wind. Renewable electricity powers our homes and businesses and we replace all the machinery with electricity-powered appliances. We change to heat pumps for heat, air conditioning, and hot water. We change to LED lighting, electric stoves and convection ovens, and electric refrigeration. Industry is served by a new green hydrogen economy. All these changes use less energy, produce far less exhaust, eliminate carbon monoxide poisoning risks, and are far less expensive. Those who install solar or wind energy systems on their own rooftops gain independence if they want it, or radically reduced costs if they stay connected to the grid, which is converted exclusively to emissions-free energy sources. In all these ways, life is actually *better*, not worse. Freedom increases. Liberty abounds. Expenses and the cost of living go down. And we cut emissions to solve climate change.

At this time, manifesting this vision isn't that difficult. We have already seen that market forces are behind a move to solar, wind, and geothermal energy. Capitalists are building factories to pump out the new devices, solar panels, wind turbines, EVs and other essentials of a new green economy. Governments are enhancing programs because they are needed, and also because of the new freedom from oil-related geopolitics. It is all good except for one thing… *time*. We simply don't have time. We don't have time for the stupid debates discussed in the previous section, and we don't have time for the arguments against renewable energy sources. This section of the book focuses on the technologies—where are we today, where we need to go, and how we meet the objections raised by sometimes well-meaning, sometimes nefarious people against the very technologies that will transform the world.

Many people say that to stop emitting carbon is simple. "Just leave it in the ground!" But they fail to acknowledge the reason that is impossible—burning carbon meets essential human needs. *You cannot just stop burning carbon without a replacement.* People in northern latitudes would freeze or starve to death. People in some southern latitudes would fry. Trade in food stuffs would be impossible. Transportation would cease. Your faucets would no longer run water and the toilets would no longer work. Cooking would be impossible.

Human society must meet these needs. The survival of society and of each of us individually is dependent on meeting these needs. The reason we do not just "stop burning stuff" is that if we do, we die. Not next year, not even next month… but *today*. Hence we know that society will continue to burn hydrocarbons until there is a better way. That better way is through technology, and we are fortunate to be on the verge of making some of the needed technologies available and ubiquitous. Others are still developing.

The good news is that we can do this. It won't come fast enough for many people and places, but it will come. And we can do it better and faster with the right help from society, government, and the private sector. What we don't have today we must research and develop. What is too expensive today must be scaled up and prices brought down. What is already viable must be adopted.

# How We Can End Climate Change

Climate change is a solvable problem. We know what we need to do. Carbon in the atmosphere needs to decline to 350 ppm according to IPCC, or even better, back to 280 ppm—closer to what it was at the dawn of the industrial revolution. All evidence today is that we are still going in the wrong direction, but that doesn't mean it needs to continue. Our situation is dire. Climate impacts are getting worse faster than expected. Average temperatures are continuing to rise and extreme weather events are becoming increasingly frequent. We need to act fast, but as nearly everyone concerned about climate would agree, we are not acting fast enough. So, what do we do? How do we solve climate change quickly?

## *How We Can Actually Solve Climate Change*

The problem is that there is too much carbon in the atmosphere. The solution is to stop putting carbon into the atmosphere and get the carbon that's there out—and do enough of both that we can return to carbon levels of 350 ppm or lower.

## *Stop Burning Fossil Fuels to Meet Our Energy Needs*

Your life, my life, and the daily life of most human beings on the planet can be powered without any emissions at all. We've been collectively lazy as we meet our needs with fossil fuels mostly because that's how we have always done it. We can easily electrify everything in our homes and offices—stoves, ovens, heating, cooling, hot water, clothes drying—all of it can be electrified. And so can our transportation, whether we used cars, bikes, motorcycles, trucks, or trains. Electric boats and planes are coming soon as well, and probably more recreational vehicles as well—ATVs, snowmobiles, and jet skis, for example. (You may, as I do, hate the idea of enabling more ATVs in the wilderness, but a lot of people don't hate them, and from a climate perspective, we have two choices—either convince them of how stupid their love of ATVs is (good luck with that) or provide them with an ATV that is free of emissions.) In other words, all of our toys, tools, appliances, and equipment will be powered by electricity.

That electricity must be emissions-free. Electrifying everything in a home, office, garage, or factory removes all emissions from the location, but it doesn't alone mean that everything is clean. The electricity must come from solar, wind, geothermal, hydropower, or nuclear, although some other non-burning generation sources such as tidal power may also work. Electrification works to eliminate emissions only to the extent that the source of the energy powering it is emissions free.

Solar, wind, and geothermal energy can be obtained individually by building a rooftop array, installing a rooftop windmill, or building a geothermal system into your home or business. Individuals who do this usually enjoy large savings over the electric company, a measure of independence, and a strong sense of having done their part to help solve climate change. Given the challenges ahead, such as extreme heatwaves and overtaxed grids, the independence provided by off grid solutions, in particular, may be highly attractive for many.

From a societal perspective, the challenge of siting emissions-free generation on the rooftops of homes and businesses is that it cannot work for everyone. Many sites are shaded, some people don't like rooftop gadgetry, some people will oppose it politically or ideologically, and it can create conflict in neighborhoods. In addition, rooftop solar requires the homeowner or building owner to opt-in, make an investment, and manage a system instead of simply paying the monthly electric bill. Many consumers will balk at this. For all these reasons, we can never achieve 100% generation from emissions-free sources mounted on rooftops. If the goal is to generate all of our energy from renewable sources, rooftop renewables can play an important part in adding to the overall solution, but they can't get us there by themselves.

While one may prefer the independence of rooftop renewables, climate change will be far more broadly affected and emissions reduced as utilities create large-scale emissions-free generation sources they can distribute over their existing grid. The reason is simple. When utilities go to non-emitting sources of electricity, *everyone* participates including climate deniers, those who can't afford to invest in solar, and the neighbors who hate the ugly panels—and participating takes no effort. All you do is pay your bill—no investment, no management, and no maintenance. Hence, even climate deniers can see no change in their lifestyle, yet still participate in cutting emissions, whether they want to or not. If your utility cuts its emissions by 20%, when you use that electricity, you have also cut your emissions by 20%. If the utility cuts by 34%, you have cuts your emissions by 34%. System changes get *everyone* on board.

This is a vision of the future than can inspire us. We can have what we need by providing it to ourselves differently than we do now. We don't have to live with less, downsize, shrink the economy, or become socialists to solve the climate crisis. Innovation and entrepreneurship can help us build a better tomorrow. And it should.

### *Scrub Fossil Fuel Emissions Where There Is No Alternative*

As much as we will try to eliminate all emissions, there are some industries that today cannot totally eliminate carbon emissions. Steel, cement, chemicals, and the fossil fuel industry itself all burn fossil fuels to manufacture their products, and cement releases carbon in the process of making lime. While it is a tempting call to arms to argue that these industries should just shut down, that is not realistic. Steel, for example, is in most of the new appliances the world needs to switch to for electrification—refrigerators, stoves, heat pumps, air conditioners, etc. Steel is also necessary for EVs, building construction, solar arrays, wind turbines, and geothermal energy. Likewise, cement is essential for the foundations of wind turbines, roads and bridges, marinas, and of course, housing. Chemicals are used in everything from circuit board manufacturing (therefore, phones, computers, and everything that has a digital anything in it) to food production (yes, even organic chemicals) to consumer products like toothpaste, lighter fluid, and hair styling products. Inks, printing, and art supplies; fishing line and poles; even alcohol manufacturers use chemicals. There's just no way around it—contemporary society needs these industries. Because the world really can't do without the products of these industries, we have to find a way to make them without emissions. This reality offers two possibilities—either invest in a new way of making them or scrub the carbon out of the process.

The first possibility is to find a new way to make steel, cement, chemicals, and fossil fuels that we do still need. Indeed, some of this is happening today but with mostly incomplete results. For example, about 30% of the world's steel is made with Electric Arc Furnaces, which is great, but the furnaces require pig iron, which is usually made only by burning coal. Other aspects of steel production can also be improved from a carbon perspective, but the displacement of massive investments in current technology makes this harder for the industry to do. It's not that these things are impossible, it's just that they will be slow to adopt without massive investment, whether by the companies or their governments.

The second approach is the most timely—creating carbon scrubbing technology and deploying it in the manufacturing processes of these and similar industries. Known as CCS (carbon capture and sequestration) or CCUS (carbon capture, utilization, and storage), carbon capture is essential to success. Indeed, an IEA report says, "Reaching net zero will be virtually impossible without CCUS." This approach is surprisingly robust in industry today with many projects being researched, developed, and tested. More importantly, industry is investing in them.

The process involves capturing carbon in the effluent stacks and then sequestering it. Some players are focused on the capture while others are on the storage. Pipelines are being invested in to move the carbon to safe storage locations. These aren't just smoke and mirrors (although there is plenty of that as well). Industry news is full of large investments in carbon capture and sequestration projects. Companies are investing in scrubbing their operations, offering sequestration services, and building pipelines to move the carbon. Here is a sampling:

- Occidental's BlueBonnet Hub is a Texas-based sequestration hub served by pipelines being built by Enterprise Product

Partners and others. It includes a lease commitment on 55,000 acres for sequestration.

- Petronas in Iraq initiates carbon capture in Iraq.
- Aker Carbon Capture is building plants in Norway and The Netherlands. It is a pure carbon business focused on CCUS — carbon capture, utilization, and storage
- Project Tundra in North Dakota CCS is a $1.5 billion project to move carbon captured from five states to a location in North Dakota for sequestration.
- Chevron and JERA sign an MOU to figure out carbon capture projects together for the US and Australia.

These investments are mostly driven by what companies can see as inevitable—new costs imposed on carbon releases by governments around the world. Carbon taxes, carbon markets, and similar mechanisms are forcing industry's hand, and while companies prefer to maintain their profits, they can also see where the world is going. They are investing billions of dollars in these facilities and systems, and when the money gets real like that, there is a reason for it. Powermag.com, an industry magazine, sees it as a booming business. Some, where they cannot scrub effectively, are buying offsets in carbon sequestration from the air, particularly with reforestation projects, and that gets us to the third approach to hitting Carbon 350.

### *Take Carbon Out of the Air*

If the goal is carbon at 350 ppm and our current level is 417 ppm, you can see the problem. Even if we suddenly stopped emitting all new carbon into the atmosphere, we would still need to find a way to remove a lot of what is already there. This is no easy task. Although the increase from 350 ppm to 417 ppm is creating havoc with earth's ecological systems, carbon is relatively diffuse in the atmosphere. This reality makes it hard to remove. We have two basic approaches—natural systems and technological systems.

### *Technology*

Technological systems for direct carbon capture from the air remain in their infancy. Universities around the world are conducting research, with one recent $1.5 million grant being awarded to the University of California-Davis. Likewise, companies are making investments in this area as well, such as Carbon Clean, which captures carbon and uses it in a new process for making green cement. Another project is Baker Hughes's recent investment with HIF Global. This project is designed to take carbon out of the air and make use of it in manufacturing synthetic fuels. Increasingly, one can imagine that as carbon comes out of the atmosphere, it could become an input for other projects—materials used in manufacturing or as a 3D printer feedstock. It could also be shipped, piped, and stored.

Companies around the world are racing to prove they can pull carbon out of the air and store it. Climeworks, a Swiss company, recently announced its new process for doing just that, and others are following suit—as many as 130 projects are in the works, though mostly at very early stages of development. Climeworks appeals to individuals who want to neutralize their so-called carbon footprint, as well as businesses who must do so to control costs. The challenge all these companies have is that most of the technology for carbon removal requires a lot of energy, and thereby creates as much or more emissions as they can capture. If the energy provided is non-emitting, such as geothermal or solar electric, for example, then there is a net gain. On the other hand, those sources could also be displacing other carbon-creating sources. As IEA says,

*The energy used to capture the CO2 will determine if and how net-negative the system is, and can also be a significant determinant of the cost per tonne of CO2 captured.*

Technological solutions are, nonetheless, developing, and we can expect that they will emerge in a highly viable way. The fact that we are not quite there today does not mean we can never get there.

### *Natural Systems*

The other option for removing carbon is natural, and the two dominant approaches are through forests and oceans. There is ongoing research regarding methods to enhance the carbon absorption capacity of the earth's oceans, as well as how to take the carbon out of the oceans to create more capacity again. Neither of these approaches, however, appears to be very practical or effective so far. So that leaves us with forests.

The climate community has been focused on forests for a long time. Forests, especially lush forests like the Amazon in South America, soak up a tremendous amount of carbon into the flora grown in the forests—through leaves and into the trunks of the trees. So as the climate crisis has worsened, more and more attention has been paid to the deforestation happening in the Amazon. And the situation has been very bad. Seventeen percent of the Amazon is gone already, and researchers say there is a tipping point at 20-25%. And the BBC reported that 2021 had the highest rate of deforestation in the Amazon of all time.

But the Amazon is not the only place where forests matter. Deforestation is a problem pretty much all around the world. The question is how to end deforestation and let the forests do their natural thing—absorb carbon to grow trees.

Activists have focused on regulation to stop deforestation, and to some extent, this works well. The recent election of Lula in Brazil, along with his pledge to stop cutting the Amazon, is a case in point. But the challenge is that a man like Bolsonaro can always be re-elected, thereby unwinding the progress. Indeed, all regulatory approaches have this challenge. Policies can be reversed, and if all else is the same, one can predict that such regulations will, in fact, be overturned once the focus on climate change has shifted. Right now, it is the most urgent issue for billions of people. In the future, other concerns will overwhelm it, especially once we seem to have it under control. Then, all of sudden, people will forget, and the policies will likely reverse.

The other method is through private contracts, most of which are monetized on carbon markets. An increasing number of companies are working with landowners to manage their lands in accordance with carbon management requirements. The best of these certify the management, much like organic farms are certified with a third-party certifier, though some have their own internal certification of compliance with the contractual obligations. Strong certifications are required for monetization, and the carbon credits are then sold to companies or individuals wishing (or required) to offset their carbon production. The firms involved in this financialization of carbon operate in various niches—some with large corporate landholders, others with small landholders; some more with the management of existing woodlands, and others more focused on reforestation.

Beyond the compliance issue, these programs have other problems—namely the greenwashing of carbon production. The first to buy these credits were those with green brands to maintain—outdoor companies, eco-tourism companies, organic companies, and so on. These companies may sincerely wish to further reduce their carbon production, but have instead bought credits, and therefore may skip the harder work of carbon reduction.

Whether the green brands skip the hard work or not, big carbon emitters will use these credits to avoid the necessary changes in their operations—at least to the point where it is cheaper to maintain operations and buy credits than it is to change how they operate. Big carbon emitters can always buy more credits, provided their main operations remain profitable. Hence, there is a sense of letting companies off the hook, by enabling them to buy carbon credits.

While individually, there may be an escapist purpose and a lack of true accountability, the overall picture is more important. If companies, for whatever reason, provide money to enhance forestation and ocean management practices, it is a good thing. Healthy forests capture carbon. So do healthy oceans. Whatever the mechanism may be, expanded management to capture carbon is essential. It eliminates the need for pipelines, sequestration locations, and capture technology since it is all there in the natural process of forest growth.

Funding vigorous forests and oceans by any effective mechanism is essential to solving climate change. Given our political-economic system, it seems natural that two forces would come into play—government regulation and a creative private sector that financializes carbon. Many people will hate to see the financialization of carbon, while others will detest regulations. The criticality of our situation, however, suggests that the means are far less important than the outcome. If these approaches can manage forests to improve carbon capture, forests can play a significant part in our overall solution.

Technologies are going to improve and other methods will be developed for taking carbon out of the air; the only question is, as with other climate solutions, will it happen fast enough? Can we reduce emissions, stop emissions, and reduced existing carbon in the atmosphere? We don't know. All we know is that we need to keep moving and move faster than we are today. Too many people have already experienced the catastrophic impact of climate change, and more are to come. We need to solve this now.

# The Answer Must Be Systemic

As I just argued, climate change is upon us, it is urgent, it is dire, and we can still take action. Indeed, we must. We all need to take action. But the drive to "do something!" has gotten people focused on the wrong things as we try to maintain choice for individuals. In fact, in most western countries climate solutions come down to a matter of personal choices, and that's the problem. Whether it is to degrow, consume less, avoid fast food restaurants, ride your bike more, become a vegan, install solar, or buy an electric vehicle (EV), we leave it up to the individual. Some people will do their part, and some will not. Hence, even the most optimistic and rosy dreams are susceptible to nonparticipation. As we saw previously, one-quarter of Americans don't even believe climate change is happening, so why would they choose to change anything at all? Plus, if the solutions are painful, that quarter of the population won't be the only ones to refuse.

Indeed, a constant refrain at the individual level is that people want to do the right thing—buy an EV, switch to electric appliances, subscribe to renewable energy—but "darn it, it is just too expensive!" Ever stop to think that is exactly the argument made by businesses—the ones so many of us want to blame? Same argument—it is too expensive. Read another away, people are saying that they refuse to participate in the solution *because other things are more important to them.* And given the fact that their participation and investment has such a minor impact on the whole problem, they are right. Many people stand ready to moralize about the decision making of other people, yet in their own lives make the same argument. But non-participation also makes success impossible.

Nonparticipation makes all the rosy predictions in these "what if" scenarios pale before the reality. Let's say you have an action that would make a 10% reduction in greenhouse gases — something like getting everyone to become a vegan. You promote it, you advocate it, and you get everyone who is going to voluntarily become a vegan to do so. Success! But as it turns out, a lot of people just like beef and chicken, so half the people don't become vegans. Therefore, total greenhouse gases are only reduced by half of the 10%, or 5%. That's only if you saturate your conversion at 50% of the people, and that is unlikely.

The same notion is true for EVs, installing solar on a home or business, signing up for wind farm subscriptions, degrowth, and any other effort. They only work if everyone complies. We need universal participation to make a difference.

### *How to Achieve Universal Participation*

We need *involuntary* participation in climate solutions. It cannot rely on people's choices anymore. The problem is too urgent and immediate.

Western societies already have involuntary participation in all kinds of things. We are compelled to work (involuntarily) so that we can participate in markets for water, food, labor, housing, energy, and all necessities of life. We must drive on a certain side of the road. For decades, people universally bought their power from their power company with no choice over how the energy was created. They participated involuntarily in the burning of coal and gas for their energy.

As climate change was debated, free market fundamentalists successfully turned the debate into options that people could adopt or not, thereby avoiding the change that was needed. Unless you volunteered to go green, the default option is your involuntary purchase of energy from the burning of coal and gas.

And yet, herein lies the model we can use to make the necessary changes just as involuntary and universal as they were with coal and gas.

### *Universal Participation Is at the Utility*

Just like coal and natural gas are the fuels utilities use to generate electricity today, wind farms and solar farms with storage, nuclear, and geothermal can be those sources tomorrow. Utility-scale renewable generating facilities with storage are large installations and operate very similarly to a coal-fired or gas-fired power plant from the standpoint of the grid and the customer. All the energy is created in the location where the facility is, then it is shipped to business and consumer users via the grid. From the user's perspective, there is absolutely no difference in how they experience the energy. You plug in whatever you are using, and it works. It's just that simple.

Except, it isn't.

It isn't that simple because a profound change has occurred out of view when the energy is coming from a renewable wind or solar farm rather than a traditional power plant — *there are no emissions*. Even more importantly, *the customer cannot make a choice* — customers use green, emission-free energy whether they want to or not.

This lack of choice means that consumers and businesses are getting green energy rather than dirty energy simply by being connected, and now it is not a decision to make, but a fact of life — just like fossil fuel burning is today. For every percentage point gain in renewable's share of utility energy production, everyone's electric carbon impact changes by that same percentage point. As such, we are no longer dependent on individuals choosing to "go solar" or choosing to consume less in a bid for a different lifestyle. The choice is made by the power company, and the faster they do it, the better.

Similar choices could be changes through manufacturing guidelines, regulations, or market forces. We already know that carmakers are retooling to make EV cars the only kind they make. As that happens, none of us will have a choice. Want a new car? It's an EV. Similarly, market forces may relegate furnaces to the ash heap, replaced by heat pumps of different kinds. Gas stoves may give way to electric stoves and ovens. The same could happen with water heaters—even the instant water heaters that are gaining popularity. One would hope that like the car manufacturers, appliance and industrial equipment manufacturers will see the same writing on the wall and move to retool, thereby leaving consumers first with better, less expensive choices than we have today, and later, with no choice at all other than the electric options. This is how everyone gets on board.

In other words, we need to stop thinking in terms of giving consumers choice and relying on them to do so. Instead we need to focus at the systems level where we can power our modern lifestyles while eliminating emissions.

## *The solution for everyone*

Utility-scale renewable energy is critical to the overall renewability and cleanliness of electricity, and it is critical for success against climate change because it involves everyone. All of us become involuntary participants in a solution. People do not need to be convinced, cajoled, or persuaded. They don't need to change their lifestyles. They can simply electrify their households and businesses, and the utilities will provide green energy that is just like the energy they buy today.

# The One Thing That Changes Everything in the Climate Fight

As we have just seen, the American system may have a lot to contribute to solving climate change. Yet those ideas are largely aspirational without participation by the greatest power on earth—big business. Big business is not solely American, but the need for business to be involved includes American business. Indeed, when the world faces a huge problem, there is no better partner to have working on the solution than American big business. The good news is that business is coming on board with climate solutions.

Since the 1970s, environmentalists have been challenged to justify "the extra costs" of doing the right thing for the environment. When that justification came on a macro scale, the argument turned to who pays. It created an uphill battle. The only businesses undertaking true environmental stewardship were those whose brands were enhanced by such actions. Each company could view it as an investment in their reputation for doing the right thing. Organic companies, outdoor companies, and selected agribusinesses tended to lead the way. Others continued with efforts to greenwash their businesses, but this again was viewed as an investment in reputation management. Sometimes it worked, and sometimes it didn't. But one thing was always clear: Doing the right thing by the environment was always more expensive.

Until now.

There has been a quiet sea change in the climate change world. So quiet, in fact, that most people have missed it. But it is there, and it is having an enormous effect.

The change is this: Over the last ten years, the cost of solar energy has dropped by about 90%. In 2010, the solar company I worked for installed solar for about $12–15 per watt. Most projects were deeply subsidized. When I left the industry in 2021, large rooftop installations were averaging $1.50 per watt, and ground-mounted systems were approaching $1 per watt.

What does that mean? According to a study by IEA, the cost of electricity from coal production in the US is $110/MWH, whereas the cost of solar is $44/MWH — 69% less cost. Similar differentials are true around the world, including in China and India.

This means that long-term planners in the public and private sector are shifting their plans for a new energy future, and projections of climate outcomes are quite different from what they were only a few years ago. Based on assumptions that coal would remain cheaper than renewables, many modelers were predicting more than 5 degrees Celsius of warming, and they were looking at catastrophic outcomes. Today, those expectations are shifting.

### *How Capitalism Is Driving the Solution*

What's important here is why the change in modeling and predictions has occurred. The cost of solar has dropped and that has made the shift to renewable electric energy the rational economic choice for virtually all businesses, governments, utilities, and organizations. Read that again. Solar and wind, in particular, are now the rational economic choice for people throughout the economy. It's not about being a do-gooder. It's not about altruism. They are not taking a *moral* stand on this, but rather a hard-nosed *financial* stance. The truth is, it is now about capitalism. Competitive price pressures are forcing the adoption of renewable energy. Capitalism, of all things, is now driving the solution.

While interest groups of various kinds are promoting their agendas for reduced consumption, spartan living, or even socialism as a response to climate change, the animal spirits of capitalism are pushing and changing economic behavior. No doubt, government policies have had an effect igniting this industry and the downward spiral of costs, but equally, no compulsory government policy could move the renewable industry forward faster than capitalism when the industry meets the need. It is astonishing to write it, but *capitalism is now on our side.*

Why might this be the case? When you make a big investment in a fossil fuel powered power plant, you have to recoup that extraordinary investment over time, but every day you run the plant, you have substantial operating costs. In economics, this cost is called the marginal cost of reproduction. In a power plant, it is defined by how much fuel and labor cost is needed to generate the next kilowatt-hour of electricity. In fossil fuel plants, that question's answer has two components, and you can see them on your bill. First, there is the standard rate you pay for electricity, say $0.06/kwh. That number is usually calculated from a combination of labor, maintenance, transmission, amortization of capital costs, and return on investment. The second number is something called the fuel cost adjustment, and that number reflects changes in fuel costs. In many places, that can be another $0.10/kwh, plus or minus a lot. For my purposes, it is not how high the fuel cost adjustment is; rather, it is that the fuel cost exists at all. The average total cost of electricity in the US is $0.166/kwh. With fuel cost equaling $0.10/kwh, you can see the impact of fuel costs.

### *Renewable Cost Advantages*

Solar's cost advantages appear in different ways. For individual businesses with available space to build it, amortized construction costs can reduce the total cost of electricity by anywhere from 40% to 80%, depending on what you pay today. Your cost per kwh goes to zero, and all you need to pay is the financing cost of building the system—no generation charge, no taxes, no fuel cost adjustment, and no additional utility fees. Add in the US tax breaks, and many businesses I worked with were seeing paybacks in as little as 4 years. (In 2026, most of the tax breaks are gone. However, the core financial advantages of renewables persists.) After that, the electricity is essentially free.

For utilities and connected consumers, the advantages are twofold. First, it is less expensive to build solar and wind than it is to build a coal or gas-fired plant today, so the first number on the bill can be lower. If you are paying $0.06/kwh today, that number can go down substantially in the future. That's because one of the big drivers of that number is the underlying amortization of the construction cost, and because solar and wind cost less to build, that number will go down. Second, there is no fuel, so that second number on consumer bills, which is a huge component of the total cost, is zero. There are no fuel costs because solar and wind don't use fuel. In other words, the marginal cost of reproduction is radically reduced and approaches zero.

Until now, part of the rhetoric on climate change has been a dire warning: "The era of cheap energy is over," people have said. As far back as 1990, a Washington Post article opined on this. People have said that we had a 200-year blowout sale on hydrocarbons, and now we have to pay the price. No doubt, there is a price to pay, and we will do it in the form of carbon mitigation strategies and climate catastrophes. But it does not need to come in the form of higher energy prices going forward. The new reduced cost of producing renewable energy means that the era of cheap energy is not over. Indeed, we are moving into an era of even cheaper energy that is free of the carbon waste causing climate change.

### *Is Capitalism's Shift Good or Bad?*

Is this good? It depends on your point of view. For those looking to promote degrowth and an end to consumerism, cheap renewable energy is likely the death knell to their movements. When energy is cheap, like it was with fossil fuels, it is difficult to stimulate the kind of austerity that the degrowth and anti-consumption movements would need. The movements anticipated, however, that climate change would require massive changes in how capitalist society operates. Many commenters favor that kind of austere world and they may well have been correct that such a world would be required had solar and wind energy not dropped so much in cost. Now, however, a low-cost, energy-driven economy can continue.

From a different point of view, this is great news. It means life can go on without major disruptions in lifestyle, standards of living in many countries can continue to improve, and we can achieve the emission reduction requirements at the same time. Rather than disrupting everything about how we live, all we need to do is switch the source of the energy we use. We need to build and buy electric vehicles, electric appliances, and electric heating products that will use that source of energy. In addition, our industrial processes and methods need to electrify and go renewable. As electrification becomes ubiquitous, costs will go down, not up. We can rely on that to ensure that the ecosystem required for a new energy infrastructure will arise because the logic of capitalism supports it.

# Conversion Infrastructure — The Key to Systemic Change

### *Get Consumers on Board*

We need a solution that can be accepted by people as an improvement in their lives at all scales—individuals, families, towns, neighborhoods, states, and the world. Technology becomes widely adopted only when nearly everyone views it as an improvement, at least over time. The iPod's "a thousand songs in your pocket" provided people with things they wanted, and therefore became nearly universally embraced. The telephone, electricity, and Internet service were similarly embraced. The automobile and the personal computer were, too. All of them made life better for millions and millions of people. That's why they embraced them.

To solve climate change, we will need the same wide-scale adoption driven by improvement in people's lives. We need better ways to meet their needs than the current method of burning stuff. We get transportation, heat, air conditioning, cooking, hot water, and more by burning stuff. We burn gasoline, natural gas, propane, diesel, oil, coal, wood, and garbage to generate heat and create energy. Either the heat applies to our indoor environs directly, such as with furnaces and hot water heaters, or it comes to us in the form of electricity after the burning fuel boils water to create steam that turns the great turbines of electric generation plants. The resulting electricity is then distributed over the grid and converted again to power for cars, heaters, stoves, etc.

### *Electricity—The Key to Conversion Adoption*

All of the heat and energy we need, in all the applications mentioned as well as industrial applications, can come from electricity. Why is this important? Because electricity can be generated from burning stuff, but it can also come from conversion — of sunlight, wind, tides, rivers, and more. Conversion technologies do not burn anything so there is no carbon or other gases emitted from the generation of energy. They simply convert natural energy like sunlight or wind into electricity. No burning at all.

Conversion technologies are scaling up. Solar panels are improving, wind turbines are improving, and we are finding ways to use them in industrial as well as consumer scale applications. There are challenges, to be sure, particularly with storing or distributing electricity so it can be used at times when the sun is not out or the wind is not blowing. But increasingly, these are technical questions that we can solve.

The solution to climate change, then, is to continue expanding the conversion energy economy and to build infrastructure and products that will use it. Electric vehicles that work better than our current vehicles in every way and every application are essential. Electric furnaces, electric heat pumps, electric water heaters, and electric industrial machinery. The infrastructure to use them is just as important — EV charging stations that work as well as gasoline fill-ups, electric stoves that are better than gas, and the storage and distribution technology that will ensure reliable supply at all times — all these are essential to achieving the mass-scale adoption that is needed.

Why do so many people in the world drive cars? Because driving cars made their lives better. There was no big decision to convert to cars. There was no authoritarian saying you needed to drive cars. Rather, people bought and used their cars because they could get places faster and more comfortably, thereby improving their lives.

In countries with high emissions, cars were adopted incrementally and society developed to support that — improved roads, better bridges, and easier life than managing horses. Adoption of automobile technology became nearly universal in such societies, and the horse and buggy disappeared.

Whether you think this change is good or bad is beside the point — this transition happened organically as people assessed a new technology, decided it was affordable enough to make an overall improvement in their lives, and therefore adopted it nearly universally. This is the kind of change needed to solve climate change

— carbon-eliminating technological improvements that make life better so that the technology can be nearly universally adopted.

*What technologies make life better?*

First and foremost, electric vehicles (EVs). EVs are an improvement on the internal combustion engine in almost every way — they are faster, simpler, and less expensive to maintain and repair (with approximately 90% fewer parts). The range on a charge is getting better and it is not hard to imagine that new battery technology will eventually extend the range. EVs don't use gas, so the price of gasoline becomes irrelevant — EV owners have no budget for gasoline at all. All of these features are personal benefits for owners.

Oh, and EVs create no emissions.

Second, consider rooftop solar. A solar array on one's home or business can often supply all of the building's energy needs. Initial investments are usually paid back in 3–4 years, and the electricity is free after that for the next 35 years or so. A stable energy supply is maintained by providing excess energy to the grid and taking it from the grid when you need it — kind of like a virtual battery — or by purchasing batteries to store the electricity and provide it back when needed. Current battery options increase the cost substantially, however, so most people connect to the grid. Electric bills disappear with rooftop solar. Building owners profit handsomely when they go solar.

Oh, and rooftop solar provides electricity without emissions.

Third, consider electric appliances. Water heaters, stoves, air conditioning, dryers, and heat pumps, for example, can all be electric, and electric versions are simpler and safer than their gas-burning counterparts. Electric versions do not have carbon monoxide risks, they don't require venting of fumes, and they don't require gas lines throughout the building. Hence, they are easier to install and maintain, and especially when coupled with rooftop solar, electric appliances are far less expensive to operate.

Oh, and since they don't burn anything, electric appliances are emissions-free.

Do you see my point yet?

Emissions-free solutions exist in current technologies, but more importantly, the reasons to adopt them are self-evident — not because they are emissions-free, but because these technologies make life better for most people. When you own an EV, a rooftop solar array, and electric appliances, your life will be far better than it is today — less expensive with better performance.

As a result, people will embrace and adopt these technologies. They are exactly what we need. As adoption increases, society will begin to adapt, just as it did when cars became attractive for most people to own. Charging station infrastructure, new building codes to include solar or electric appliances, and the elimination of natural gas infrastructure are all likely to emerge. As they do, adoption will increase again, thereby creating a positive vortex in society.

### *Our challenge*

The challenge is to get that positive vortex spinning. EVs are still relatively rare and charging stations even moreso. The same is true for rooftop solar and many electric appliances. If there is a role for government, it is not to impose a degrowth strategy or build a space colony. Government can help by reducing the cost of adoption to make all these benefits obvious. In the US, we have tried to do that but fallen short. Tax credits on EVs and rooftop solar, for example, help wealthy people obtain these benefits, but they are not available to people who can't use the tax credits. President Biden's initiative to build out an EV charging station infrastructure has good intentions, but the stations feature an out-of-date technology. These efforts have failed to get the positive vortex spinning.

We don't have time to mess around. If the government can reduce the cost of entry for everyone, the remaining benefits are so strong that mass adoption will follow. If there is an advocacy position to take, it should be reducing that cost—not "leave the oil in the ground." Government helps make people's lives better by enabling them to get EVs, rooftop solar, and electric appliances, and in so doing it creates a solution to climate change.

This improvement in life gives technology an edge over all other solutions — people want these changes. As a result, the technologies can be universally adopted, and they will permanently phase out older methods and replace them with emissions-free technologies.

### *Get Industrial Producers on Board*

Consumer choices are only one part of the equation, however. The production side also has to be turned into something that workers, managers, and investors favor and benefit from — especially those who are deep into careers, ways of life, and investments in oil exploration, refining facilities, and so on. As much as climate activists may not care for these players, they must be included in the "everyone has to benefit" pool I mentioned previously. It has to be good for everyone. So, how do we do this?

The country needs to invest green energy infrastructure into the areas currently serving and dominated by the oil, gas, and fossil fuel industries. West Virginia can become home to carbon capture and the wages can outpace that of the coal miners while investment returns surpass those of mine owners. Louisiana and Texas can focus refinery activities on plastics feedstocks, including those needed for 3D printing, and the remainder of their industrial capacity can be converted to conversion equipment and renewable energy production. North Dakota's fields hosting oil rigs can also host enormous solar arrays and wind farms. Plus, the oil wells can be used for geothermal purposes. Where the capacity and expertise can be developed within the existing companies present in these locations, utilize that. Where it can't, buy them out.

Industrial processes, however, may require alternatives to electricity as an energy source. This is where hydrogen comes in. Hydrogen can create the heat industrial processes need—it is the same fuel used to send rockets into orbit and beyond—without producing carbon. The first step is to create hydrogen through a process known as electrolysis. Electricity is used to split water atoms into oxygen and hydrogen. As it turns out, if that electricity comes from fossil fuels, there is no net reduction in emissions. If it is produced by solar, wind, or another clean source, there are no emissions and the hydrogen is clean.

Once hydrogen is produced, it has two potential ways of being used—fuel cells or direct combustion. A fuel cell uses the hydrogen, air, and the anode and cathode of the cell to separate electrons from the protons of the hydrogen, thereby creating electricity. Applications for this kind of cell have been discussed for vehicles or airplanes. The more industrial use of hydrogen would be through direct combustion to get the heat or power that such combustion produces. Yes, this is technically "burning" something, but it is emissions-free burning since there is no carbon and no carbon dioxide is produced.

The problem with hydrogen is that it is the smallest molecule in the universe so it is difficult to contain. Pipelines created for natural gas, for example, cannot be easily re-used for hydrogen because the hydrogen would leak. Engineers are working to solve this problem, but it remains a challenge at this writing.

We can see the potential of conversion infrastructure. It is there and evident. Other countries, especially China and Germany, can see it. China especially is building it. But here in the US, there remain arguments against renewables, especially solar and wind. To that, we turn next.

# Debunking the Arguments Against Solar and Wind Energy

Child labor. Rare metals are needed. Poor battery technology. Baseload power requirements that renewables can't meet. End-of-life recycling. Effects on pollinators. Dependence on China. The arguments against wind and solar go on and on. To which I can only say… *Really?*

The planet is burning up. Millions already suffer from climate change, and millions — even billions more — will do so as well. We have an obvious solution to emissions that will make people's lives better. And all they have to say is, "Nyet"?

## *Consider the Arguments*

Renewable energy is produced without direct emissions. In today's world, emissions are required to build these systems because we have to burn fossil fuels while mining the materials, manufacturing the parts, and assembling them. That's obvious. It is also obvious that as electric vehicles and tools become ubiquitous and we use renewable electricity to operate them, these emissions will be reduced. But we have to get from here to there.

The use of child labor to build solar cells is an abomination. So is the use of child labor to make shoes, bricks, fireworks, and carpets, which are some of the most common industries where child labor is used. The problem is child labor, not the product. No one is stopping the use of shoes, bricks, or carpets because some manufacturers use child labor. Instead, they are trying to certify that suppliers are not engaging this horrific practice. Likewise, we should stop the use of child labor for the production of solar cells, not stop using solar cells.

There are obvious technical limitations to solar and wind energy because we need energy 24/7, but the wind doesn't always blow and the sun doesn't always shine. We certainly must develop ways to store the energy that is generated with renewables to achieve an excellent long-term solution, and those methods of storage are being developed. The deployment of solar and wind projects today pushes the technology along and creates incentives to develop it further tomorrow.

End-of-life recycling is always an issue with any development, be it automobiles, buildings, iPhones, computers, or forks and spoons. Yet we do not stop producing or buying such items because we haven't solved that problem.

Wildlife, particularly birds and pollinators, are affected by wind turbines and solar arrays respectively. One study in Canada, however, showed that while wind turbines will eventually kill about 233,000 birds a year, windows in Canada kill about 25 million birds per year—or one hundred times more. Yet, no one is up in arms about windows.

Worse, climate change itself will lead to the extinction of 2/3 of all species of North American birds. Read that carefully. Not just 2/3 of the birds, but 2/3 of the species that will never be heard from again. That is, 100% of the birds of 2/3 of all species will die due to climate change. Wind turbines will kill some birds, but they won't make species go extinct; in fact, they can help reverse climate change and preserve all bird species.

Off shore windmills are also hitting snags with local groups claiming that a recent rash of dead whales in the northeastern US is probably caused by disturbances offshore windmills create in the waters. These arguments are being made as scientists warn of the total collapse of the north Atlantic current known as AMOC—Atlantic Meridional Overturning Circulation. A recent study published in Nature included these excerpts from the abstract:

*The Atlantic meridional overturning circulation (AMOC) is a major tipping element in the climate system and a future collapse would have severe impacts on the climate in the North Atlantic region… Tipping to an undesired state in the climate is, however, a growing concern with increasing greenhouse gas concentrations. Predictions based on observations rely on detecting early-warning signals, primarily an increase in variance (loss of resilience) and increased autocorrelation (critical slowing down), which have recently been reported for the AMOC. Here we provide statistical significance and data-driven estimators for the time of tipping. We estimate a collapse of the AMOC to occur around mid-century under the current scenario of future emissions.*

A systemic shutdown of the Atlantic Ocean's circulation system would be devastating to all wildlife in the ocean, not just the handful of whales that have washed up recently on the shores of New England. In fact, those whale deaths may already be caused not by windmills, but by warming waters and the collapsing ecosystem that we cannot see, but which is happening anyway. Yet, these oppositions are slowing progress. Soon, there won't be any whales to even think about if we don't solve climate change.

All of these objections are real issues, and all of them need to be addressed. But they are raised as objections to renewable projects, with the result of delaying or killing the projects.

Here's the reality: If we do not solve climate change, nothing else matters. The planet will become uninhabitable. Desperate people will migrate in an attempt to survive. Global population will shrink in the most horrific way. Many species will go extinct. Children will labor to save their lives in all-new ways, making today's child labor issues look tame. No one will care about end-of-life recycling because we won't outlive the technology.

Sometimes these objections are followed by absolute silence — no proposal, no ideas. There is no alternative agenda, just a desire to stop progress for whatever reason. Sometimes it is NIMBY (Not In My BackYard) concerns. Sometimes it is *climate denial.* Sometimes it is nothing more than smug righteousness. In these cases, we are dealing with opponents to solutions, not alternative solutions or priorities. Every movement has opponents, and climate is no exception. The mistake we may make, however, is to assume that the silence indicates a friendly position. It does not. Silence is a purposeful end game to stop progress. Where opponents offer no acceptable alternative, it makes things worse.

### *Arguments Against Solar*

Given our climate emergency, it is a puzzle to me as to why so many environment-oriented commenters argue against the feasibility of renewable energy — especially solar. "More energy goes into making solar panels than they will ever produce," I heard several times. "They require rare-earth materials," was another. "You can't build industrial-scale solar," was a third.

I'll take the last one first. I hate to disagree, but as the former VP of Sales for a solar power company focused on industrial sales, I can tell you with absolute certainty that solar can be done and is being done — on industrial rooftops at industrial scale. I sold three of the five largest rooftop projects in the State of Minnesota to date — all of them over 1 MW in size, and each preventing industrial quantities of greenhouse gases.

Second, solar panels are estimated to require about 600–800 kWh of electricity to produce them, and that amount of energy is produced by a panel in 2–4 years, depending on your location and orientation to the sun. Good panels are warranted for 25 years, and most last 40 years or more. A 10–1 or 20–1 energy payback is very workable, and we know that the technology will improve. What's more, we can produce tomorrow's panels with yesterday's solar array installation, and when that occurs, there is no longer any emissions created by solar panel production. It's as clean as you can get.

With regard to rare-earth materials, current solar panel technology requires them. That technology will improve, however, and the requirement for such materials will decrease over time. In the meantime, solar panel construction will require mining operations to acquire them, along with the environmental and cultural destruction that goes with mining. Then again, continued oil & coal production is already doing the same thing. Hence, we have two bad options — continued destruction via oil production or similar destruction via mining — but one leads to a solution whereas the other does not. As we move over to a solar renewable energy base, resolving the need for such materials is a top priority.

# The Lynchpin to Electrification: Battery Storage at Scale

In addition to the various objections to wind and solar energy we have seen, deployment has been limited by our inability to adequately store the energy. Wind and solar energy tend to overproduce when the sun shines or the wind blows, but at night or on breezeless days, they produce nothing. If you need energy on those breezeless or cloudy days, or if you need energy at night, it's not available. This causes problems for consumers and users, and also for the grid, which needs a relatively constant load to function properly. Hence, we have the debate about storage.

Although there are many viable alternatives for storing energy, batteries have emerged at the head of research and investment. Today, there is a race to improve battery technology to make it more capable of storing and deploying energy to meet our needs. Many promising technologies are out there, but only the lithium-ion batteries we are all familiar with have become standardized so far.

The primary argument against batteries is based on current chemistry, materials availability, and rare earth metals needed. All lithium-ion batteries, whether in phones, watches, laptops, EVs, or utility-scale battery banks, include cobalt, and over three quarters of the cobalt mined in the world today comes from Democratic Republic of Congo. Unsurprisingly, the "gold rush" has led to fighting, killing, and refugees. There are at least 120 armed militias fighting for control in the provinces where the cobalt is mined. Millions of people have been displaced. Cobalt is functioning in this country has oil has functioned elsewhere.

Although this situation is concerning, it is no reason to stop development of solar and wind energy systems. In the first place, most applications are grid-tied, which is to say that they have no batteries at all. Most industrial-scale solar and wind projects do not use batteries. They are either utility scale and feed the energy into the grid, or the buildings on which wind turbines or panel arrays are mounted use the energy, and any excess is returned to the grid. Once on the grid, the utility directs the energy to where it is needed. There is no need for batteries so long as these renewable sources are supplementary to the other sources on the grid, such as coal, natural gas and nuclear energy.

Batteries are rarely used at industrial scale because they are not yet cost-effective, and the critics are correct that this has to do with the material required, as well as the chemistry characteristics. But battery research is ongoing, and new developments will radically change battery feasibility in the future.

The industry's incremental improvements are outlined by the Rocky Mountain Institute (RMI) which published a report on batteries in 2019 that describes the central role of lithium-ion batteries today and the possibilities for new batteries in the future. All of them are moving past cobalt, and most are moving past lithium. According to RMI, these batteries are likely to be commercially viable by 2025. A few examples from RMI's report include:

- Zinc Batteries: Different kinds are developing, including zinc-anode batteries and zinc alkaline batteries. The first will create a less expensive battery with an improved life cycle.
- Flow Batteries: Early flow batteries are beating out lithium-ion in certain use cases already.
- High-Temperature Batteries: RMI says: "Liquid-metal batteries could provide low-cost, long-duration grid balancing based on their safety and long life cycles and preference for active cycling, similar to traditional generators."

### *Thermal Batteries*

To understand the viability of such systems, it pays to look at the business world. Wherever people are investing, there is usually something worth discovering, and there is a lot of investment currently in high-temperature batteries, also known as thermal batteries. Antora Energy, which is backed by Bill Gates and other investors has raised over $80 million and appears to be the leader in terms of bringing a real product to market. Several other players, including Thermal Battery Corp. are scaling up as well. When money like that is pouring in, you usually have a cost-reducing technology that can be adopted quickly.

CNBC described the process: *"Electricity from wind and solar are run through coils, like in a toaster, to heat solid, well-insulated blocks of carbon to over 3,000 degrees Fahrenheit, storing a tremendous amount of energy. The blocks can be shipped cold in a battery module to factories."*

Thermal batteries are especially suited to the storage of large amounts of electricity and will likely be used by large industrial facilities in addition to utility scale projects. If they work, these batteries will solve a huge challenge for the industrial and utility markets.

### *Seawater Batteries*

Other innovations appear in projects like IBM's seawater battery. Here is how they describe it in this IBM Research blog post:

"Using three new and different proprietary materials, which have never before been recorded as being combined in a battery, our team at IBM Research has discovered chemistry for a new battery which does not use heavy metals or other substances with sourcing concerns.

"The materials for this battery are able to be extracted from seawater, laying the groundwork for less invasive sourcing techniques than current material mining methods."

### *Wood Based Batteries*

The Scandinavian company Northvolt is working on a new battery based on wood, which they claim will be the world's most sustainable battery. Nearly everything in it is recycled materials, including cobalt and manganese, and wood is its core. It turns out that there is carbon in wood that can be used as the anode, or what they call the lignode. The company is also powered by renewable energy and seeks to decrease the carbon footprint per cell produced by 90% by 2030. Assuming the batteries are just as effective as lithium-ion and competitively priced or cheaper, this is a game changer.

### *Tesla's Four Million Mile Battery*

For obvious reasons, Tesla is very interested in battery technology. They created a Tesla Advanced Battery Research division in 2016, which is now developing a lithium-air battery that far outlasts the current generation of lithium-ion batteries. The key to this battery, however, is the development of a single-crystal cathode to replace the polycrystalline cathodes used in today's batteries. Polycrystalline cathodes are filled with fissures that enable the early breakdown of the cathode, whereas single-crystal cathodes have almost no fissures and therefore prevent breakdown. That's how their life is extended. In ongoing, day-and-night, charge and discharge tests run over the last four years, the battery is still at 95% capacity, which implies a twenty-fold increase in the life expectancy of the battery.

The point is that research is continuing and until the battery questions are solved, we can continue developing the solar infrastructure and reducing carbon emissions — even without batteries.

### *There Is Another Way—Hydrogen*

In this TED talk podcast, Denmark's climate minister, Dan Jorgenson, describes how Denmark is going to end the country's oil industry altogether by 2050. This is a big deal since Denmark is the EU's biggest oil producer. But what I appreciate is how Jorgenson's plan focuses the minds of millions of people toward a societal level goal. That's the leadership we need. Here's how they plan to do it.

First, build as much off-shore wind power as they can — about 10 gigawatts, which is far more than the country needs. Denmark was the birthplace of wind-to-electric power generation, and it will continue to lead the way with applications of wind power.

Second, use the excess electricity in the electrolysis process to create hydrogen, which can be stored and used in fuel cells. In electrolysis, ectricity is used to split hydrogen and oxygen from water. Since the electricity is all excess wind-generated electricity, hydrogen essentially stores the energy.

Third, replace internal combustion engines of all kinds with hydrogen fuel cells or direct combustion — including airplanes, trucks, and industrial machinery. Although automobiles can operate by storing electricity in onboard batteries, today's technology isn't feasible to use batteries or electricity for everything. The batteries would be too heavy for the airplanes to lift, for example, and you may not be able to concentrate the electric energy enough for some industrial applications. Hydrogen fuel cells bridge that gap. This step allows the use of hydrogen to power these machines.

Fourth, use hydrogen fuel cells to power these machines that electricity cannot power. In effect, using the hydrogen in the fuel cells releases the wind-generated stored energy. No emissions from wind-sourced energy, and no emissions from the hydrogen fuel cell. Water and heat are the only byproducts of the chemical reaction that releases the energy.

In essence, the excess electricity generated by wind power will be stored as hydrogen to be used at a later time. It is brilliant.

Jorgenson's plan acknowledges that for some applications, a fuel-based system will better meet our needs than an electric-battery-based system. Hydrogen allows us to use fuel without burning anything. Rather, it is just a chemical reaction that releases energy.

The point should be clear: Researchers and engineers are working to solve the electricity storage problem. Whether it is batteries for cars or hydrogen storage, enormous efforts are in process and the storage challenges will soon be solved.

# The Cost of Carbon Can Now Be Calculated

As hard as people are working to solve climate change with solar, wind, geothermal, hydroelectric, and battery technologies, it seems prudent to recognize that we are going to fall short. Climate impacts are coming on faster than scientists originally predicted, and they are getting bigger than we expected sooner. As a result, stopping emissions is only one part of the solution. As we have seen previously in the book, carbon is now at 417 ppm in the atmosphere. Our target is 350 ppm as defined by the UN IPCC. The pre-industrial era saw 280 ppm. Clearly, we are going to have to get rid of some that is already there. That's' where mitigation and carbon capture technology come in the play.

According to reports, the Swiss company Climeworks has become the first company to remove carbon from the air and deposit it underground where it will turn into rock, permanently locking it away. The implications go way beyond just getting the carbon out of the air. I'll explain in a minute.

As with all new technologies, the first pass is very expensive and the challenges of scale are critical. These are normal problems of technological development. First, cell phones were impossible, then they weren't. Then, cell phones were way too expensive, then they weren't. Then, the network couldn't handle them and scaling could not work, then people figured it out. We are at a very normal, predictable place in the development, deployment, and commercialization of new technology with carbon capture, and the reality of what Climeworks has done should be celebrated. Over time, this too will scale and costs will come down. They always do, and this is no different.

### *No Free Pass for Fossil Fuels*

One of the biggest concerns environmentalists articulate is that with carbon capture technology, there is less incentive to switch from fossil fuels to other ways of creating energy. On the surface, that seems true, but as the old saying goes, the cat is already out of the bag. There are more powerful incentives to move away from fossil fuels than climate Armageddon. Namely, *renewable energy is now less expensive than fossil fuel-sourced energy.*

If you don't think that is a big deal, consider this. You are the manager of an electric utility that needs to replace aging generation capacity. That is, it needs to build a new power plant. You need to get board approval for a massive capital expenditure. All in and fully amortized, the fossil fuel plant will generate electricity at $0.04/kwh for thirty years. The solar installation to generate the same amount of electricity will do so for $0.02/kwh for the same thirty years. From a purely business perspective, how do you make the argument for the fossil fuel plant? "We want to build in a 100% higher cost structure so we can lose money, pollute more, and use the oldest technology available." Good luck with that.

Five to seven years ago, this argument would have been completely reversed. You would have to argue for solar by claiming it was "the right thing to do," but you were swimming uphill against bottom-line economics for the business. Today, it is the opposite. The burden of proof is now on the fossil fuel option. How do you justify the extra expenditure, the unpredictability of fuel access, and the ongoing cost of fossil fuel-based plants?

### *Now We Know the Cost of Carbon*

Massive carbon extraction at scale could offset the environmental damage of fossil fuels, but it won't reverse this inflection point on the relative cost of renewables. In fact, it may do the exact opposite. A company like Climeworks can now monetize carbon removal, which is to say, it can assign a dollar cost to such removal. Whether that removal is paid for by individuals, companies, or governments, the cost is there for everyone to see. We can expect a battle royale over the question of who should pay the costs of that removal. Should it be the producer of the fuel or the entity that burns it? Should it be the product builder, like car companies? Or the refiner of the fuels? Or should the government handle it as a byproduct of modern economies, thereby letting taxpayers pick up the bill? Who pays is to be determined. Climeworks and other companies working in this market, however, establish a market price for the removal of carbon. There is no longer any need for government to set such a price or to establish trading markets in carbon credits. We now know the actual cost of carbon.

Through one mechanism or another, this cost of carbon removal will be added to the cost of living in modern society. Most likely, it will be a mix. Producers of fossil fuels will pay some, burners, including consumers, will pay some, and the government will pay some. In economic terms, this technology enables society to remove carbon pollution as an externality. Suddenly, carbon can be accounted for and costed, and therefore participants in the economy can be held accountable through more accurate and honest costing.

## *Only One Part of the Solution*

Don't get me wrong… carbon capture is only one part of the multi-part solution we need. Scaled up, this can buy us some desperately needed time. It can help us avoid the idiocy of atmospheric engineering while enabling the world to continue converting to emission-free energy sources. That conversion requires not only the expansion of renewable electricity generation but also the adoption of all-electric appliances, from cars to heating/air conditioning to hot water to trucks and airplanes. That conversion is happening, but it will take time.

The electric-renewable route I see coming requires that new infrastructure be built. EVs need charging station infrastructure. All-electric appliances need new installation skills for HVAC people, electricians, and others. Housing codes will need to change. Rooftop solar servicing will be required.

Other options also require infrastructure that takes time. Many people advocate more total use of mass transit in our urban centers, while others advocate local infrastructure for growing our own food, producing to meet our own needs, and building community. Still others advocate a radical reduction in consumption and a degrowing economy. Whichever of these is your preferred solution, they all have one thing in common — they need time to take root. The infrastructure behind them needs to be built out. Carbon capture at scale can buy us more time to build it.

When all is said and done, we are going to need emission-free energy sources, reduced consumption, and carbon mitigation to tackle climate change. The breakthrough by Climeworks is a huge step in the right direction.

SECTION 3

# *The Way Forward*

Given our situation, almost everyone I talk with asks the same question: What can I do? It is a good question that often gets a wrong answer. Instead of the alternative lifestyles and the "doing without" that many people recommend, the real answer lies in the guidance from productivity gurus, coaches, and business managers—we need to adopt strategies that will have the greatest effect for the lowest effort. It is no longer a question of "Can I just do with a little bit less?" but rather, "How can I make an outsized contribution to the solutions we all need?"

The first question leads one to small things that are easy to do. "You can do x, and y and z and it isn't even that hard." Okay, you do them, and little has changed. The second question, on the other hand, asks for a bigger commitment to the biggest challenge of our time. We are going to explore how to have the biggest impact possible in this closing section. But you should get ready… it just might change your life.

Ultimately, as we have seen throughout this book, our system for creating energy has to change. We need a system that converts energy to a form we can use instead of one that burns stuff and creates carbon emissions along with the energy. The technologies are at hand, but they are not yet built into the system. The system is crucial because only when the system changes will everyone come on board. Just as we have all been on board for centuries with burning fossil fuels because it was the only system available, when that system no longer burns fossil fuels yet meets our day to day needs, we will again all be on board with climate solutions—not because we agree or choose but because it is the best way to live.

What we can all do, then, is find a way to contribute to that systematizing of climate solutions. This last section of the book focuses on the personal choices that can make a huge difference. We are looking for impacts that go beyond any measure of personal impact like your emissions or carbon footprint. Those actions require time and energy, but they will only affect one person. They are important to do, but we can do better. That's what I am pointing to here.

Let's get started.

# Make Money and Make a Difference

Unless you are a climate denier, you are probably wondering what can be done to make a difference on climate. Answers abound. Reduce your carbon footprint. Get an EV. Go solar. Ride your bike. Others get focused on activism—protest, make your voice heard, smear paint on a famous piece of art. All of these are add-ons to our regular lives. But what if you could *make your life about solving climate change?*

Ever since I was a kid in the 1970s, I can remember hearing lectures, radio interviews, television discussions, and later Youtube interviews and podcasts in which someone asks a speaker: "What can we do? What can we as average consumers do to help?" Most of the time, the answer was "write your congressional representatives." But in the environmental movement, there was more—conserve energy by turning down your thermostat, recycle, and live with minimum impact. Those were lifestyle changes you could make. In the climate debate, of course, it has become a list of actions you can take to reduce your carbon footprint.

There is a problem with the question, however. When we ask "What can we as average consumers do?" we limit the range of our potential impact to the sidelights just mentioned because we are only asking about our role "as consumers." We are also *producers*—that is, employees, workers, entrepreneurs, and investors. We commit our best energy and most of our waking hours to our work. These roles provide us with our biggest opportunity to contribute to solutions with the highest impact on emissions and the economy. It's not about what we can do *as consumers*, but rather, what we can do *as human beings*.

## *Think High-Impact and High-Leverage*

We all know we need to wake up and do the right thing. But the right thing has less to do with one's personal carbon footprint and more to do with participating in high-impact, high-leverage solutions. What would it mean to have high impact and high leverage in your climate solution activities? Greta Thunberg is an example of someone who has had an outsized impact on raising awareness. There are others. Love him or hate him, Elon Musk has driven auto companies to go electric through innovative competition. Because Tesla proved the viability of electric vehicles, virtually all car companies have EV programs. Some people are inventing and bringing to market perovskite solar cells, which improve efficiency and produce 50% more electricity in the same amount of roof or ground space. People doing battery research and bringing real solutions to market are another example.

It's not that we can all be Greta Thunberg or Elon Musk, nor can we all be Bill Gates, who is investing in and making possible a multitude of climate-related projects, or Bill McKibben, who is effectively getting people focused on systemic economic and political change throughout the world. We can't be them, but we can be engaged in high-impact, high-leverage activities.

What is that high-impact, high-leverage choice? For most of us, the single biggest way to multiply our impact on climate is to change our jobs. Engineers can get into solar and wind development, battery research, EV and other electric appliance development, grid engineering, and so on. Marketers can help people understand the opportunity, and sales reps can go sell the stuff. Finance and administration types can commit their efforts to companies and organizations creating solutions. Writers and speakers can keep calling people to a higher vision of what they can contribute. Entrepreneurs can go looking for great ideas and pull together people like Gates to help fund them. Installers and electricians can build rooftop solar, install heat pumps, and replace non-electric appliances with the electric appliances of the future. That single change will unleash more creative energy toward a solution than any other single thing a person can do.

Indeed, the only way to unleash the creativity and power of the human spirit in the service of climate solutions is to ***make the solutions the focus of our employment***. All of us put our *best* energy into our jobs and we put *most* of our energy into our jobs. If your job isn't part of making the necessary changes in technology happen, then you are not contributing much to the solutions we need. The world needs your genius, your ability, and your unique abilities to solve this problem. If you want to make a real commitment to climate change, get a new job in which you can commit forty, fifty, or sixty hours a week to climate solutions. Focus on real solutions that reduce atmospheric carbon.

### *Climate Jobs Are Plentiful and They Pay Well*

At the same time, we are running short of our goals due to a lack of workers. PV Magazine recently laid out the situation:

*"In order to reach SEIA's Solar+ Decade target of solar making up 30% of all U.S. electricity generation by 2030, SEIA estimates that the solar and storage industries will need an estimated 800,000 new workers for a total workforce of more than 1 million. DOE estimates that as many as 1.5 million workers will be needed by 2035 to achieve the Biden-Harris Administration's decarbonization goals."*

Those are big numbers and lots of jobs. When I worked in the industry, we had projects that could not get built because we didn't have the people needed. And this is just one part of the new green workforce. HVAC and all home appliances need to be converted and newly installed. Car chargers need to be installed. Workers are needed in new lithium mines, battery factories, EV factories, and more.

It is no longer a matter of sacrifice for a greater good. Collectively, we need to work as fast as possible to become an emissions-free economy because that is the only way to stop emissions and eventually reverse climate change. Even if you are marginal on climate, the money and benefits offered in these salaries is serious, and there is opportunity for everyone.

Filling the labor gap in the industries solving our emissions challenge is, for most people, the single best way to magnify your impact. Take your unique skill set, find a business or organization that is making a difference, and commit your time, energy, and skills to that endeavor. The solutions require all capabilities used in operating a business, from product development to HR, accounting, sales, customer training, marketing, and service. There is a role in green business for anyone who chooses that today.

But, don't do it out of the goodness of your heart. Green jobs are paying great and the lack of workers is slowing down the conversion of the economy to emissions-free. According to salary.com, *"The average Solar Installer salary in the United States is **$78,173** as of May 25, 2023, but the salary range typically falls between **$64,651** and **$96,691**."*

New industries rise and fall only a few times through one's life, even though they do so repeatedly through history. Those who catch these industries as they rise always make all the money. From lumber to railroads, steel, automobiles, chemicals, petroleum, software, and the Internet, those who get in early reap the rewards. This is true as employees as well as stockholders. Today, many of the best jobs in the world are with Google, Apple, and Microsoft, but the time to get in was years ago. What is on the new leading edge? Two things—the green economy and AI. The two are complementary. We think of the green economy as EVs, solar, wind, grid engineering, battery engineering and manufacturing, carbon capture, and more. AI is operating autonomous EVs and managing the grid, and will assist in operating the green economy of the future. And because these are new industries, they are paying big salaries to attract people to build them out.

Recent searches on Indeed.com show the following openings:

- 4,320 battery engineer jobs
- 155 climate activist jobs
- 100 wind engineer jobs
- 733 jobs as PV solar system designer
- 7,473 solar sales representative
- 3,085 solar marketing jobs
- 29,121 HVAC installer jobs

The high impact industries to enter now focus on new sources of energy and the new tools and products that run on them. New sources of energy include solar, wind, geothermal, battery research, electric grid management, and more. New tools and products are everything electric—heat pumps, electric vehicles, electric water heaters, electric trucks, and more. Bring your engineering, marketing, business, HR, and other skills to those efforts because that's what is needed and that's what will make a difference.

As you can see, this is a calling to a bigger commitment. No, being vegan won't solve climate change. Taking the bus or commuting on a bike won't either. Do those things because they make a small contribution, but I am asking you to consider a real commitment—*your career*. If you care and you are intelligent, you won't settle for the small impact. Change your job and you will change the world.

# A High-Impact Personal Climate Action that Really Matters: Buy an EV

While we just talked about the producer side of the equation, there are big impacts to prioritize as a consumer, too. What is the best thing to do as a consumer? Buy an EV. Many people will decry this as just more wanton consumerism, but it is not. Yes, EVs currently require mining for lithium and rare minerals. Yes, you still need steel or plastic for car body parts. Yes, there is a carbon cost, at least today, to building an EV. But an EV has something none of the "consume less" strategies have—direct impact on innovation that will require real qualitative change throughout the energy economy. Here's what I mean.

First of all, when you buy an EV, you are giving a big middle finger to the oil companies. Why? Because 45% of crude oil is refined into gasoline. As an EV owner, you become one more consumer who will no longer buy gasoline. This is not temporary. This is not simply to pressure the oil companies. This is to disconnect from the oil companies altogether. You no longer need them and you will no longer buy their primary product… *ever again.*

Second, when overall demand for electricity increases, production must increase to meet it, otherwise you get brownouts or rolling blackouts. Electricity producers do not want to build ahead of time for fear of creating stranded, unused assets. Increased demand, driven by your EV, will motivate utilities to build more production. When they see the need, they will look at costs, see that solar and wind are far less expensive than coal or natural gas, and they will opt to build solar and wind. Increasing demand means investment and construction of renewable generation—market forces will require it. Every EV in consumer's hands adds to this overall demand that the utilities must respond to, and with solar and wind power so cheap, it is the only choice they can make to meet demand. Hence, EVs promote the construction of renewable energy production.

The third leverage point for buying an EV is that your purchase drives R&D on battery technology. As good as wind and solar are for generating electricity, their intermittency requires improvements in battery storage for them to take higher percentages of market share. Many companies are active in R&D on battery technology. This work will underpin the development of new storage technologies not only for EVs but also for utility-scale storage. Your EV purchase supports that kind of investment as well.

Finally, EVs pull energy off of the grid. As such, they incrementally increase the required capacity of the grid. So as you and your neighbors adopt EVs, electric utilities can see the increased demand and in addition to building generating capacity, they will also build grid capacity. Electric utilities will not invest in building out a bigger, better grid if the demand isn't there—again, that would create stranded assets. By buying an EV and charging it on the grid, your consumer action signals to the electric company that demand is increasing and it will support the investment.

Grid improvement is critical to building the energy infrastructure of the future. The current grid is woefully inadequate in many ways. In fact, it is one of the number one impediments to a systemically renewable future. Here is a chart from Power Magazine showing the number of projects in the queue for grid connection by year and energy source.

Note that nearly all of the new projects are renewable, but they are held up because of inadequate grid capacity. According to this study, the backlog amounts to 2,000 gigawatts of capacity, whereas the total installed capacity of the current grid is 1,250 gigawatts. The backlog is approaching double of all existing US electricity production powered by fossil fuels, nuclear, and renewables combined. The investments are ready but the grid is not. The connections for these projects can only be approved when the grid is upgraded to connect with them, and that means electric utilities need to see demand. Your EV purchase will add to the demand.

This is why a new EV purchase may be the most impactful thing you can do for climate solutions. *It's not just the vehicle, it's the leverage you get in changing the energy economy.* You get a better, less expensive car to drive—that's for sure. But you also get off of oil altogether, help enhance the grid, help drive necessary battery R&D, and motivate electric companies to keep building renewable energy sources. *None of this happens through conservation or consumption reduction.* In other words, just like the original automobile industry changed everything in the US economy and around the world, the EV industry can do the same thing. The more of us that buy EVs, the more quickly that necessary change will occur.

# Are EVs Better on Emissions?

EVs may have the biggest leverage available for a consumer action on climate, but there is a lot of controversy about whether or not they are all they are cracked up to be on carbon footprint. The dubiousness of that measure notwithstanding, I decided to take a look when inspired by a reader who asked three key questions:

*1) What is the carbon footprint of a mid-size gasoline vehicle that averages 25 miles per gallon and is driven 15,000 miles per year in ppm?*

*2) What is the carbon footprint of an equivalent-sized electric vehicle driven 15,000 miles per year based on the daily charge rate required in kilowatt-hours converted to ppm produced by the most commonly used type of power plant?*

*If the ppm value is too coarse for the measurement, please use the finer version of the same measurement. I am not familiar with ppm or its meaning.*

*3) What is today's most commonly used method to produce electric power for our grids?*

First, let's get the terminology right because it can be confusing. Ppm is "parts per million" and is a measure of concentration of carbon in the atmosphere. The higher the number, the worse our situation has become regarding climate change. In pre-industrial times we were at about 280 ppm; today we are near 420 ppm. More importantly to the reader's question, ppm is not a measure that can be assigned to a car, a plane, or a power plant. It is a measure of total carbon in the atmosphere.

Instead of ppm, therefore, we measure the carbon coming out of the tailpipe of a car or a smokestack. It is measured by weight—either pounds, tons, grams, or kilograms. That is a relatively easy measure to obtain. So, according to shrinkthatfootprint.com,

*"Fuel combustion emissions can be calculated from an actual fuel economy, as we know around 8.8 kg CO2e is produced using each US gallon of fuel. So if you know your car averages 25 MPG then you can divide 8.8 by 25 to give 351 g CO2/mile for fuel combustion emissions."*

Oddly enough, that is a direct answer to the first question: CO2 is produced at 8.8 kg per gallon of gasoline burned, or 351 g CO2/mile if the internal combustion engine (ICE) vehicle gets 25 mpg. If you drive 15,000 miles in a year, you produce 132,000 kg CO2.

To make a direct comparison to EVs, it is easy to do—EVs have no tailpipe, so they have no direct emissions at all. One mile, 15,000 miles, or 150,000 miles, it doesn't matter—no emissions because there is no tailpipe.

So the first comparison is tailpipe emissions:

- ICE emits 132,000 kg CO2 for 15,000 miles driven
- EV emits zero kg of CO2 for 15,000 miles driven

But this is deceiving. People are quick to point out that while the EV does not emit carbon, much of the electricity used to power the EV may come from power plants that burn coal or natural gas, both of which do have emissions. They are right about this. Essentially, EVs on the same grid we have today mostly just shifts the emission from the tailpipe to the smokestack. But is there any savings?

To calculate this, one needs to know how much of your electricity comes from what source. The website reneweconomy.com.au spells it out:

*"The carbon intensity of electricity varies greatly depending on fuel source.*

*"As a rough guide coal has a carbon intensity of about 1,000g CO2/kWh, oil is 800g CO2/kWh, natural gas is around 500g CO2/kWh, while nuclear, hydro, wind and solar are all less than 50 g CO2/kWh."*

In other words, fossil fuel-based electricity emits ten to twenty times more carbon than nuclear and the renewables, so the source of the electricity really does make an enormous difference. Let's calculate how large that difference is.

We can start with how many kwh are needed to drive a mile. Just as mpg efficiency is different from one ICE vehicle to another, the same is true for EVs. One source put the average as 0.346 kwh per mile using 2020-2022 models and data. But a quick glance at this chart shows that the range can be from 0.24 kwh/mile to 0.87 kwh/mile—either 50% less than the average or as much as 150% more. The type of EV you buy has a big impact on the amount of electricity needed, and thereby on overall CO2 emissions.

The other obvious questions is this: What is the mix of electricity sources you are using? The extremes are 100% coal versus 100%

renewable. Most people only get 100% renewable if they are producing their own power from rooftop solar or have subscribed all of their energy to a community solar or community wind project. If your energy comes from the grid, it is most likely a mix.

This reality probably drives the reader's third question about the most common source of power for the electric grid. As an answer, the US Energy Information Administration says this:

*In 2022, about 4,243 billion kilowatthours (kWh) (or about 4.24 trillion kWh) of electricity were generated at utility-scale electricity generation facilities in the United States.1 About 60% of this electricity generation was from fossil fuels—coal, natural gas, petroleum, and other gases. About 18% was from nuclear energy, and about 22% was from renewable energy sources.*

As with efficiencies of ICE vehicles and EVs, these numbers are averages, and the range is large. This chart from Wikipedia shows that the range is from 99.7% renewable sources in Vermont to 3.2% renewable sources in Delaware. Hence, the answer to the reader's question is extremely dependent on where you live and where the energy on your grid comes from. It is an important question because it affects the answers to his questions about EVs.

### *A 300-Mile Trip Example*

Let's consider an example—a 300 mile long trip. We will have to use these averages and acknowledge that individual cases could vary greatly.

In an ICE vehicle that produces 351 g of CO2 per mile driven, multiply 300 miles times the 351 g per mile and you get 105,300g of CO2. Keep in mind that the 351 g CO2 per mile is based on 25 mpg fuel efficiency. While 25 mpg is a reasonable average, higher mpg would reduce the amount of CO2 per mile, and lower mpg will increase it.

To take the same trip in an EV, the average amount of electricity cited above is 0.346 kwh/mile, which is 103.8 kwh for the 300 mile trip. If all of that electricity comes from coal, you will produce 103,800g CO2 taking that trip (103.8 kwh multiplied by 1,000 g CO2 per kwh). For non-burning sources of electricity, you will produce

5,190g CO2 (103.8 kwh multiplied by 50 g CO2 per kwh). In most cases, the driver will be somewhere in between those extremes.

So in this analysis using averages for a 300 mile trip, there is little difference between EVs powered by coal electricity (103,800 g CO2) and ICEs powered by gasoline (105,300 g CO2) in terms of carbon. EV produced CO2 drops as non-burning sources of electricity increase in the mix used to charge the EV.

But this is not the end of the story. If we are looking upstream at fuel sources for electricity, we should for ICE vehicles as well. The gasoline burned in the ICE car also has a carbon cost when it is refined from crude oil into gasoline. This carbon cost has been calculated at 1.12 kg CO2 per gallon of gasoline (2.4589 pounds of CO2 per gallon). So, divide 1,112g CO2 by 25 miles and we get an additional emission calculation of 44.5 g CO2 per mile. On a 300 mile trip, this adds 13,350 g of additional CO2 emission (300 miles multiplied by 44.5 g). The puts the ICE vehicle at 118,650 g CO2 for the 300 mile trip.

To sum up this comparison so far for a 300 mile trip:

- ICE vehicle making 25 mpg produces 118,650 g CO2
- EV running on electricity from coal produces 103,800 g CO2
- EV running on non-burning energy produces less than 5,190 g CO2.

In other words an average EV will save about 15% in emissions versus the average ICE, *even if all the electricity comes from coal.* The same vehicle will save over 95% of emissions if all the electricity comes from non-burning sources.

### *Carbon Cost for Building the Vehicles*

Critics are always quick to point out that these are only operating emissions, and to get to the actual *carbon footprint*, you have to look at what it takes to build the car. This, however, is where it gets extremely murky.

For ICE vehicles, several items must be considered. There are the remaining emissions from producing gasoline that I haven't counted in the refining process noted above—petroleum exploration and discovery, production (pumping), shipping, and then distributing to customers, the factory that assembles the cars, the steel, the parts, the plastic in the parts, the glass, etc. All of these have some impact on carbon footprint, and their actual level depends on the efficiency of the supplier and whether or not they are using clean energy. When you think about assigning a portion of each of these to a product, it makes one's head spin. Mike Berners-Lee who, as we saw earlier, wrote the quintessential effort to define and measure carbon footprint, *How Bad Are Bananas?*, essentially threw his arms up in the air: *"The carbon footprint, as I have defined it, is* **the** *climate change metric that we need to be looking at. The problem is that it is also impossible to measure."*

The same problem haunts EVs. Part of the upstream carbon is the same—there are the various assembly factories and part production factories, similar as exist for ICE vehicles. Instead of oil refining and distribution, however, the EV carbon footprint must include the battery production process, as well as obtaining and refining materials such as cobalt and lithium. The carbon impacts of batteries are always included in the calculation for EV manufacturing, but I have never seen the upstream inputs from oil production included for ICE vehicle's carbon footprint. Nonetheless, the general consensus is that EVs are more carbon intensive to build than ICEs due to battery production. *"However, the initial CO2 footprint associated with EV production is dwarfed when compared to the operational CO2 footprint of ICE vehicles. The initial carbon deficit from manufacturing only takes several thousand vehicle-miles/year of ownership to overcome and depending on how "green" your energy grid is, EV carbon parity with ICE vehicles can be reached as quickly as six months."*

Here's the kicker. Just as an EV's *operating emissions* depend on the source of electricity used for charging the battery, so does the calculation of *production emissions* for both ICEs and EVs, so the calculation of carbon footprint could be off by as much as *twentyfold.* Shipping, refining, mining, oil production, and much more in the process involve energy, and the source of that energy greatly affects the total carbon footprint.

For this reason, any listing of the so-called "carbon footprint" for building a vehicle, whether ICE or EV, is highly questionable. Inherently, such an estimate requires assumptions to be made. Not only might those assumptions reflect bias, they could also be wrong or very easily change, especially as the energy input mix changes. Consumers have no way of knowing. The best estimates are that within six months to two years, any increase in carbon from EV production when compared to ICE vehicle production is made up for in its operation. Once you hit that point, you are rapidly contributing to a reduction in overall CO2 for a given amount of driving.

### *The Upshot*

The way I look at it is this. The idea of calculating the entire carbon footprint of a product, especially one so complex as an automobile, is a meaningless exercise. EVs and ICEs are roughly similar in terms of the carbon emitted in building them, and in any case reach parity after only a short time driving the EV. ICE vehicles support the fossil fuel industry—remember that 45% of oil is burned as gasoline—and so require exploration, production, refining, and all the greenhouse gases that go with them. EVs, on the other hand, have cobalt, lithium, and other materials that require mining, processing, and refining. Carbon produced to make EV and ICE vehicles is probably fairly similar using today's technology, but EVs have one huge advantage—they can get better. Battery technology is likely to go beyond lithium and cobalt as new research discoveries are made and new promising technologies come online, and as they do, the carbon footprint of EV production will shrink significantly.

For operating the vehicles, as we have seen, EVs have an advantage even if all the electricity to charge them comes from coal—they are about 15% better. As the electricity to charge them is renewably sourced, however, the EV number drops precipitously and can exceed a 95% reduction. Our best chance to solve climate change is to get our energy from non-burning sources and to use it intelligently. Even if your new EV is today drawing most of its recharge from coal-based electricity, you are still improving carbon emissions, and as your utility converts to non-burning sources, your impact will increase. So on an individual level, too, it is best to get an EV.

# How to Compare Emissions Between Your Car and an EV

*Compare Your Current Gas Vehicle to an EV*

It is all well and good to calculate comparisons using averages, but given the wide variation in the range of values involved, the averages don't always apply. In a given situation, what looks like a big carbon benefit for EVs might actually be the reverse—it depends on variables like the efficiency of the ICE car you are replacing, the efficiency of the EV, and the sources of your electricity. It seemed necessary to set up a way to compare these based on the data that are specific to your situation.

The other question critical for consumers is this: "Does it cost more?" Again, the answer is specific to a situation. The electricity to power EVs is measured in kilowatt-hours (kwh) and the cost per kwh varies widely depending on where you live, whether or not you have your own renewable source of electricity and the local cost of charging stations if you need to use them. It also depends on the price of gasoline, which can also vary widely by region, and from week to week or month to month. So these variables need to be accounted for in a cost calculation.

To take a look at the effect of these variables on a case-by-case basis, I developed a spreadsheet that enables the user to input the relevant data specific to their situation, and then see how the two vehicles compare. You can use it to compare annual usage or even for something as small as a 300-mile trip. You just need to enter a few data points, and the spreadsheet will calculate your comparisons. I'll show you how to get a copy of the spreadsheet in a moment. First, I want to identify the inputs you will need and clarify the assumptions used in it.

**The Inputs You Will Need**

Three key inputs are required: The average miles per gallon (MPG) you get in your retiring vehicle, the average kwh/mile for your new EV, and the percentage mix of your sources of electricity—coal, oil, natural gas, or non-emitting sources like nuclear and renewables. When you enter these figures, you will get the carbon comparisons based on the assumptions outlined below.

For operating cost comparisons, you will need to estimate your cost per gallon of gasoline for the ICE vehicle. Usually, the best way to do this is to consider the preceding six months in your area and use a price that is close to what you have experienced.

To calculate the cost per kwh for your electricity, however, you should pull together a year's worth of your electric bills so that you can enter both your monthly total cost and your kwh usage, which will help determine your cost/kwh. Although electric bills often list a $/kwh cost, the additions and variations on the bill add up, so the easiest way to estimate total cost is to divide total kwh usage by the total monthly bill in most cases. Just enter the proper numbers into the spreadsheet, and it will calculate it for you.

**Assumptions on Carbon Emitted from an ICE Vehicle**

According to many sources cited in the preceding chapter, CO2 is produced at 8,800 g (8.8 kg) per gallon of gasoline burned. In addition, CO2 is emitted during the refining of gasoline at a rate of 1,112 g (1.1 kg) of CO2 per gallon, for a total of 9,912 g (9.9 kg) of CO2 per gallon of gas burned. These are the assumptions built into the spreadsheet on ICE vehicles. Using your average miles per gallon, the spreadsheet calculates the carbon per mile driven.

These two sources are the primary contributors to operating emissions for an ICE vehicle. Although theoretically, you could include all the other carbon costs associated with gasoline refining, including oil production, transport, exploration, and so on, visibility to real data on these gets increasingly difficult and the contribution to the overall picture gets smaller and smaller. The spreadsheet uses only the two inputs mentioned above to calculate the carbon emissions of ICE vehicles.

**Assumptions on Carbon Emitted from EV**

EV emissions are dependent on two key factors as well: efficiency and electricity source. ICE vehicles measure efficiency in terms of MPG, whereas EVs measure it in terms of kwh/mile. Currently, EVs have a wide range of efficiency ratings, from a low of about 0.24 kwh/mile to a high of 0.87 kwh/mile. The actual rating of your EV will have a decisive impact on the comparison. So, you need to know the efficiency rating of your EV to figure this out. The EV manufacturer has a number for your vehicle, but like with MPG ratings, it will depend ultimately on how you drive and your local driving conditions.

A separate set of assumptions is made about carbon emissions based on electricity sources. As I stated in the preceding chapter:

*As a rough guide coal has a carbon intensity of about 1,000g CO2/kWh, oil is 800g CO2/kWh, natural gas is around 500g CO2/kWh, while nuclear, hydro, wind, and solar are all less than 50 g CO2/kWh.*

In other words, electricity from natural gas emits ten times as much carbon as nuclear and renewables, and electricity from coal emits a whopping twenty times more carbon than nuclear and renewables. Hence, the higher percentage of your electricity that comes from coal, the higher the total EV emissions produced by driving your vehicle will be. In most cases, your local utility is probably the best source for finding out what percentage of your electricity comes from what source.

**How to Get the EV-ICE Cost and Carbon Comparison Spreadsheet**

To get the free spreadsheet, go here: https://anthonysignorelli.substack.com/p/ev-vs-ice-comparison-calculator.

## *Replace Your Car with an Electric Vehicle—to Save Money!*

Faster, zippy, more fun, and pretty darned cool. That's what everyone is saying about EVs. We've just seen how EVs usually, but not always, lead to a major reduction in carbon emissions from your personal travel activities, especially so the more your electricity comes from non-burning sources. They also help build out a new system of electric generation based largely on renewables. But how do they work financially?

Some EVs are pricey--there is no question about that. But lower end and used vehicles are pricing in at or near the price levels of similar used ICE vehicles. The savings will depend on the type of car, efficiencies, and local costs for gasoline and electricity.

For example, I drive an older model minivan. It gets about 18 miles to a gallon. I spend about $350 per month on gasoline and have a payment of $157 per month for a total cost of $507. I can purchase a recent model EV hybrid SUV that gets 52 miles to a gallon for a $3,000 down payment and payments of about $300 per month. My new gas bill will be about $121 per month (at 52 miles to the gallon), so my total cost is $421 per month compared to the $507 today. That's a savings of $86 per month. Since this is a traditional hybrid EV in this example, the regular maintenance expenses are assumed to be the same. I get a better car, reduce my carbon emissions by 65%, and save $86 a month in the process. See what I mean?

I could also exchange the car for a 2020 all electric SUV for about $2,500 down and $362 per month in payments. My fuel cost goes to zero, but I would have about $43/mo in electricity costs at $0.15/kwh. Total cost for payment and fuel would be $405/month, a savings of $105/month. But there is an additional savings on oil changes since EVs don't need oil changes. Kelley Blue Book identifies the average cost as $35-75 per change. I figure another $45 a month on average. So by trading in for an electric SUV that meets my needs, I can save $150 a month. The emissions I am responsible for will depend on the mix of my electricity sources.

These are two examples, and not all situations are the same—not by a long shot. The prices of gasoline and electricity are crucial, as are the efficiency ratings of the vehicles. That's why I created the spreadsheet. The point is that in many situations, converting to an EV will save you a lot of money, and it is almost always better for the environment and a help in ending climate change.

# High-Impact Climate Choices That Save You Money

For decades, people have faced a choice between their money and doing the right thing by the environment, and climate change was no different. The question was: Would I pay a little more to do the right thing? Many did and many did not. "It's too expensive. I can't afford it." These reasons came from businesses as well as individual consumers. Today, that has all changed.

We've already seen that there are very good, high-paying jobs in the industries that will eliminate carbon and solve climate change. We've also seen the power of switching to an EV. But what else can you do to save money and help solve climate change at the same time? Here's a list of considerations and their potential impact.

### *Save 10% by Sourcing Your Electricity from Renewables*

Perhaps the simplest and easiest step anyone can take is to source your electricity with a solar or wind farm subscription at a reduced price. If it is available where you live, there is no reason to not do this. It won't take more than an hour, and you will save money, reduce carbon, and make electricity more reliable.

The opportunity to subscribe is not available in all areas of the country, and in some places it is offered but you must pay a premium price, not a discount. Electric rate payers everywhere should at least look into this to see if it is a possibility. Many will be happily surprised.

### *Convert Your Furnace/AC to an Air Source Heat Pump and Save BIG*

If an air source heat pump is right for you, installing one can create a big monetary savings, especially if you need to replace your furnace or AC anyway. How do you know if it is right for you? As I see it, there are two primary use case scenarios, both based on your climate situation.

Heat pumps are very efficient for heating and cooling a home. They work as a heater in the cool weather and as an air conditioner in the heat—simply reverse the flow and you have AC. Heat pumps use less than half the electricity of electric baseboard heat and are also more efficient than air conditioners.

Historically, electric heat has been more expensive than natural gas in nearly all cases, but that is changing. Today, in seven of the fifty states in the US, heating with electricity through baseboard heat, for example, is less expensive than gas. Air source heat pumps, however, use about half the electricity that baseboard heat requires, so in nearly all cases, heating with an air source heat pump will be less expensive, assuming it will work in your area. The biggest concern about air source heat pumps is the effective operating temperature range required. They are good at heating or cooling at moderate temperatures—say from about 0°F to about 95°F. When ambient temperatures get outside those ranges, the efficiency drops, and too far out of those ranges and the air sourced heat pump may not work at all, although ground source heat pumps will. If you live in an area where the temperature reliably stays within that range, air sourced heat pumps will save you a lot of money and work year round. If temperatures reliably get colder than that, you may need a supplemental heating system. In my case, I use supplemental electric space heaters when needed, but one could also use a wall-mounted propane heater, a wood burning stove, or the original gas furnace as a back-up.

For heating alone, a heat pump will always save you electricity and cost over baseboard heating, but the comparison to natural gas heating is more difficult. It depends on gas and electricity prices and efficiency of the units being compared, in addition to the temperature range of your area. I found this handy calculator that can give you a good idea based on your particular circumstances.

Operating costs, however, are not the only consideration. The purchase and installation of a heat pump will cost less than the purchase and installation of a furnace and air conditioner. Heat pump purchase and installation in almost all circumstances is less expensive than a furnace and air conditioner purchase and installation. Furnace and central AC can easily be $8,000-10,000, whereas heat pumps are usually one-half to two-thirds of that price. Actual numbers depend on the size of your home and where you live. Specific numbers depend on the types and sizes of the units involved, but heat pumps are nearly always a cost savings over traditional heating/cooling appliance combinations.

### *Build Rooftop or Ground-Mounted Solar*

I used to sell rooftop solar systems and I always thought it was one of the best value propositions I had ever seen in sales. You can build it with the cash you have, earn a payback in a few years and enjoy free electricity for the rest of the system's life—usually 30-40 years. Or you can finance it, thereby transferring your current electric bill into a loan payment usually at a savings of 10-50%, and usually without a down payment. Either way, the savings are substantial. The actual amounts, however, depend on where you live, your current cost of electricity, and the incentives available.

To me, either one of those outcomes are substantial, but there are a few things to be aware of. First, if you are seeking to be completely off grid, the numbers are not as good. That's because you also need to install a battery system and those are still fairly expensive. For this reason, most people build grid-tied systems which do not store the excess energy but rather deliver it back to the grid in a kind of virtual storage, usually getting a one-to-one credit for the energy sent back.

Second, you do not want to build a solar array on a roof that needs to be replaced any time soon. That's because to replace the roof you will need to remove the array, replace the roof, and then replace the array—a fairly expensive process. On the other hand, if you build an array over a new roof, the array will protect that roof and extend its life. If you are building a new roof or have a metal roof, solar is a fantastic addition to your carbon reduction efforts, and it will save you more money than perhaps any other single thing you can do.

Other than that, there are few downsides. Thirty years of free electricity on an investment that paid back in about five years is compelling no matter where you are.

# The Leverage of Politics

Ok, so you are going along with me so far. You are saving money by electrifying and going to renewable energy. Great. You have your EV. Great. You managed your career to something that can have a much bigger impact. That's perfect. And now you ask, what else can I do?

First and foremost—tell your friends what you have done. Share your savings, your improved life, and your story with them. Provide information so they can benefit, save money, and be part of the solution as well.

Beyond that, the most important thing we need to change the system is political. Politics matter because they change the *economics* for businesses and corporations, without whom we cannot reduce atmospheric carbon to 350 ppm. Corporations have two keys. First, some of them are enormous emitters and politics can change the cost of such emissions. Second, corporations can make system-level changes that affect everyone without us having to think about it. Hence, there are two primary thrusts of political action that will have a big impact: carbon pricing and emissions-free generation requirements for electric utilities.

## *Carbon Pricing Changes the Game*

Carbon pricing is critical because some industries that meet real needs, such as cement, chemicals, steel, and others, cannot completely clean up their production methods with current technology, yet they must stop emitting carbon if we are to succeed. If cement were a country, as one commentator said, it would be the biggest emitter of carbon in the world. Fossil fuel refining is also a huge emitter. So long as carbon emissions are free, these industries have no incentive to clean themselves up, and in fact, the intrinsic laws of capitalism compel them to manufacture at the lowest possible cost. So if emissions are free, they need to emit instead of manage emissions.

But if emissions are costly, the companies in these industries must assess the cost-benefit analysis of managing versus emitting.

Most of them know that carbon pricing is coming and they are already starting to invest in emissions reduction strategies. Our role as citizens is to keep the pressure on them and advocate for policies that ensure a high price for carbon. If we maintain that, the innovative spirits of corporate capitalism will find a solution even if they can't go to emissions-free electricity for all their power. When it comes to making money, corporations always figure it out, and carbon pricing changes how they do that figuring.

*Renewable Requirements for Utilities*

The other policy area needing a lot of attention is getting utilities to generate electricity with renewable systems. Solar is now the cheapest way to generate new electricity, beating out every other method. Solar is cheaper to build and cheaper to operate. It can be made better yet with proper public policy—incentives for entering the renewables market, renewable requirements for operating in a particular state or jurisdiction, ensuring transmission line construction needed to carry the electricity, and so on. Unfortunately, many states are going in the opposite direction, mistakenly thinking that the old way is the cheap way, and thereby making it harder to build renewables, sometimes in states where it would be especially easy to use solar or wind.

The reason the utility target is so important is simple—whatever percent of the power the utility generates is renewable, *all of its customers are the same percentage renewable.* If the utility generates 50% of its electricity from emission-free sources, then 50% of all energy used by all customers is emissions-free—regardless of their attitude toward climate change. The utility mix applies to ALL users because that is what is coming in on their powerline. That's leverage. That's impact. And that will make a difference.

These two areas can be accelerated by good public policy. So for people looking to take political action, that are two of the most important areas to focus on. Well directed activism has its place.

# Boycotts, Lifestyle, and Other Low Impact Carbon Footprint Reduction Strategies

As we saw in the first part of the book, individual action without leverage or without collective action is minimally impactful in the face of the challenges that confront us. While some people can change their job and purchase an EV for maximum impact, other people live in various situations where minimal impact may be the only thing available to them. If this is you, you will want to do what you can. The returns get lower for the amount of energy you are putting into it, but this can still be worthwhile—even if for no other reason than that you can feel you are doing everything you possibly can. After changing jobs, buying EVs, and the electrification strategies discussed previously, the political strategies of the preceding chapter have the next best impact. Then come these lifestyle decisions with all their baggage.

## *General Lifestyle Considerations*

Many people want to change their lifestyle. Consume less, become a vegan, buy your food locally and organically, or maybe even grow your own in a garden. Ride a bike. Take public transportation instead of a car. Choose your diet to lower your carbon footprint. All these are fine ideas. The science is out, however, on their actual impact on carbon, even though many advocates will insist these are the most impactful things you can do. The problem, as we have seen earlier in the book, is that carbon footprint cannot be measured accurately, even within an order of magnitude because it can't account for all kinds of upstream decisions that have a big impact. If you can't measure it, you can't possibly know the impact your actions are actually having. It is easy to tell how much carbon is no longer coming out of your chimney when you stop burning natural gas; it is much more difficult to know if your apple from New Zealand is more or less carbon emitting than the chicken grown next door. And even if you know the difference, how much does it matter?

Lifestyle choices can be good for many reasons—health and wellness, less cost, more enjoyment. But as far as carbon is concerned, you as a consumer cannot tell the difference between a high carbon footprint product and an alternative low carbon footprint product. If you want to change your lifestyle, go ahead and do it. Good mindfulness will probably yield a change in your total carbon impact, but you won't be able to measure it. Nonetheless, you'll feel better and there is likely some kind of small positive impact.

But let's get specific. Let's review a few personal actions people undertake believing they are having a big impact, but let's also be serious about how big an impact they can actually have—especially when compared to the high leverage actions discussed earlier.

*Refrain from Air Travel*

If you were able to organize a boycott to stop all air travel in the world whatsoever and ground all airplanes permanently, you would reduce carbon emissions by about 3%. That's something, but your chances of succeeding in such a boycott are very small. It will take a lot of energy, and even if you win, we still have 97% of emissions to go.

Instead, many people content themselves with their own choice not to fly—effectively, a boycott of one—and feel better because they believe they have reduced their personal carbon footprint in a meaningful way. For US airlines in 2022, their share of the 3% of emissions came from carrying 853 million passengers. Even if you assign all air travel emissions to US airlines, the effect of your boycott is 1/853,000,000 times 0.03. It is an infinitesimally small number. There is also no leverage. Despite this reality, many people feel that shunning air travel is an important step, and if you are one of them, by all means do it. But let's recognize that this is not a high leverage area likely to lead to significant success.

Because air travel is also largely a personal choice, it has also become a flash point between people. Those sacrificing the opportunity to travel often criticize those who choose to travel, and this adds to the already acrimonious tone underlying the climate debate. It is another caustic addition to the fraying social fabric. At a time when we need to solve climate change to save the human species and biodiversity on Earth, we need to find ways to work together rather than ways to despise each other. Contempt for others solves nothing. We need to become better versions of ourselves.

### *Sell Your Car and Walk, Bike, or Use Mass Transit*

Another personal responsibility decision focuses on the use of a car, and sadly, this decision also frequently results in contempt for others. The refusal to use a car has many benefits to those who choose it—you save money, you will be more healthy, and the tailpipe emissions from your car will go to zero if you never drive it. There can be real freedom in this change of lifestyle, as well as a sense that you are contributing something substantial by assuming a different lifestyle. And you are—at least in terms of your own personal contribution to climate change. On the other hand, there is little or no leverage. Your car is one of 1.4 billion cars in the world.

People often magnify the importance of this approach by saying that "if everyone stopped driving, we would eliminate 35% of emissions." The problem is that most people who drive regularly have no real choice in the matter and a quick society-level transition from cars to public transportation is impossible. We simply do not have the infrastructure for such a change; it would need to be built, and that is a multi-decade effort which many people will reject anyway. Solutions that most people won't do are not solutions, they are pipedreams.

What's not a pipedream is the transition to electric and clean hydrogen powered vehicles, as we have seen. At this time, it appears that electric vehicles (EVs) will dominate the automobile market and replace cars with internal combustion engines (ICEs), while clean hydrogen may come to dominate the heavy truck world. Much depends on technology and infrastructure development, but all of that is well under way. People who have access to the use of vehicles are not going to give them up, especially if they are living in societies organized around the use of such vehicles. Likewise, we must have a means for transporting goods of all kinds across our countries, and while rail will help with some of that, you still need to get truckloads to stores and, increasingly, to individual homes.

Bicycles and mass transit may work for many people and help them reduce their so-called personal carbon footprint, and if you are in that situation, by all means utilize it if it works for you. As a prescription to solve climate change, however, this strategy is insufficient. It just isn't impactful enough.

### *Refrain from Eating Meat*

As we saw earlier in the book, the veganism debate as it pertains to climate change is fraught with all kinds of agendas, ideologies, and misinformation. I discussed most of this in section one. Still, many people feel that this is a way to do something about climate change. Most of the benefits from veganism as far as climate change is concerned have to do with the notion that by not eating the meat, you are slowing the deforestation of the Amazon. But as with shunning air travel, the impact is only meaningful with massive collective action. Your single-handed action is meaningful to you, but its global impact is limited since you are only one of 8 billion people! If everyone stopped eating meat, that could be helpful, but that appears to have little or no possibility of happening. Plus, despite the outrageous claims by the World Watch Institute that over 50% of emission come from animal agriculture, nearly all credible sources put the number at more like 5%. Hence, you have the same problem as air travel does. Get everyone to comply with an enormous effort and significant backlash and you still have the other 95% of global emissions to deal with.

### *Consume Less*

Consuming less is a way to reduce the amount of energy used to enliven your life. This can be eating less, traveling less, commuting less, watching less TV or playing fewer video games. It can mean turning down your thermostat and taking fewer trips. It can mean borrowing books, not buying them, sharing tools with neighbors or friends, or any host of things that results in conserving energy. Like many strategies when it comes to climate change actions, you will also probably save money by consuming less.

Once again however, these are individual lifestyle choices that won't even register on the overall problem unless nearly everyone participates. Even then, the impact would be very small. If you equate consumption with emissions, a ten percent reduction in consumption would reduce emissions by ten percent as well. But if you break that link altogether, as we would with a transition to solar energy, for example, the amount of consumption is no longer relevant yet emissions would plummet. This is why, as noted in the preceding chapters, it is far more important to obtain your energy from emissions-free sources than it is to conserve or reduce consumption. It may be good to reduce usage of fossil fuel energy, for example, but how much better to have energy that is completely emissions free!

## *The Lure of Lifestyle*

If you are motivated to make a lifestyle change to help with climate change, or if that is the only avenue of change open to you, by all means, contribute what you can. People are attracted to lifestyle choices in their response to climate change partly because it is something they can control in a world that feels very uncontrollable. Let's just not stop there. Although I have written forcefully against the ideology of personal actions as the answer to climate change, they do matter. I wrote against them earlier in the book, and to some extent above, because they will never be enough to bring about the change we need, and I see many people sitting on the sidelines thinking that their lifestyle choice is enough. It is not. Each of us needs to make a personal commitment to stopping climate change, but everyone will be better off if that effort is put into high leverage, high impact actions rather than low impact, feel good actions. Given the monumental challenges ahead, we must go that way. There is no time for messing around.

# Epilogue: It's Not You, It's the System

We have seen in this book that climate change is undermining everything about modern society, not just our ecology. We have seen that solutions exist if we can just implement them. And we have seen that there are social and economic resistances built into the system that make such implementation difficult. Consumers don't have the heft to change the system upstream. Businesses are vested in business as usual, which makes it very difficult for them to be leaders. Governments are torn by political considerations and popular fears, both of which are big factors in whether or not they can maintain power and contribute to solutions. And finally, those who see systemic solutions are running into local resistance from people who want to protect their way of life and their immediate environment.

All of which leads one to ask: How in the world are we going to succeed?

First, we have to admit that we might not succeed. Climate change may carry on unmitigated, accelerating faster and faster, until whatever calamity befalls humanity and the population is reduced by so much that mother Earth is finally able to begin resetting herself. This is the snow crab solution. Wait until it gets so bad that the population sinks by 87.5%--a number that eliminates 7 billion people and reduces global population to near what it was at the dawn of the industrial revolution—around 700 million. Emissions would collapse by roughly the same amount. Earth would have to work out what is left in the atmosphere, or it could just keep cooking those remaining people until they, too, are gone. The snow crab solution becomes the dinosaur solution.

To avoid that outcome, we need a plan that unfreezes progress against carbon reduction. As we have seen, there are many ways to get distracted into other priorities or environmental problems, but such distractions will not help us. Carbon is our immediate problem, and our solutions need to address carbon. The other problems will need attention, but if we don't fix carbon, nothing else will matter. The question becomes how we do it.

The way to achieve results in organizations is to identify the big goal and get everyone focused on it, then to define actions at all levels of the organization that can be taken to help reach the goal. Then, tie that progress to specific goals and measure progress against it. Everyone focuses on their part in the plan knowing that everyone else will be held accountable for their part of the plan. For businesses, their sub-businesses, divisions, departments, offices, and individual performers all know their part in achieving the goals. In sports, each player has to play their part and perform in excellence. You can't go play another person's position because that leaves yours uncovered. Hence integration becomes coordination, and all the parts contribute to achieving the needed solution.

This is what we need to do now. Every person on earth has a unique ability and a unique position to have an impact. The more we focus on high impact actions through our work or high leverage actions like purchasing EVs, the faster we will get there. The more we focus on low impact, consumer level actions, the longer it will take. We just have to remember that high impact and high leverage are far more effective even though they are more challenging to our lives. Nonetheless, they are the only way we really solve this.

The reasons the system is so important are clear: When the system only offers clean energy because it is the best and cheapest source of electricity, every customer participates in the solution. When the only cars available are EVs, every driver is participating in the solution. When cordless electric power tools take over the market because they are better and more enjoyable to use, every homeowner becomes a participant in the solution. When airplanes fly on hydrogen, we can travel without the carbon problem.

But even more, as all these changes occur, there will be less and less need for oil refining in the system—and therefore oil exploration, production, and pipelines. Industrial processes that still emit carbon can install carbon capture technology to prevent it—provided there is a systemic cost applied to everyone. And of course, we will build out the grid that is needed to distribute electricity, or standardize on the installation of solar in all new construction as part of the new building codes and systems.

In this book, I hope that I have been able to focus attention on the real problem, the solutions to that problem, and what we can each do to contribute. I have tried to go beyond our current dogma that focuses on carbon footprint and demonstrate the amazing power of leverage and high impact—and that we can all improve our level of impact.

As we face the climate crisis, we can look at the terror that it is, but we cannot allow ourselves to be frozen by that terror. The solutions exist and they are known, and we are working to make the system work better than it does now. It is up to all of us to transform the system. You play your part, I'll play mine. As we do, there is good reason for optimism.

# Books by Anthony Signorelli

*Rooster Crows at Light from the Bombing,* (editor),
Inroads Press, 1993

*Call to Liberty: Bridging the Divide Between Liberals and Conservatives,*
Scarletta Press, 2006

*Speculations on Postcapitalism,*
Blue Harbor Press, 2017

*What Is Liberalism?*
Blue Harbor Press, 2018

*The Great Mechanism: The Power Behind the Relentless Juggernaut of Western Capitalism,*
Blue Harbor Press, 2018

*Consent Is Not Enough: What Men Need to Know in a #MeToo World,*
Blue Harbor Press, 2021

*What Is Postcapitalism? Imagining the World After Capitalism and Socialism,*
Walden Woods Press, 2025

*The Arc of Love,*
Walden Woods Press, 2026

# About the Author

Anthony Signorelli, also known as The Sailing Poet, is semi-retired as a writer after a long career in business that included a technical writing and training firm, sales consulting firm, a bookstore, and two organic farms. He is currently an co-editor of the online magazine *Inroads* and served previously as the poetry editor for Holy Cow! Press. The moniker "Sailing Poet" comes from his dual love of sailing on Lake Superior and poetry. His current work explores grief, climate change, the AI revolution, and the eloquence of the heart. You can find it at anthonysignorelli.substack.com.

www.ingramcontent.com/pod-product-compliance
Lightning Source LLC
LaVergne TN
LVHW022223200926
848481LV00006B/2098
* 9 7 9 8 9 9 5 3 4 0 6 1 4 *